PIMLICO

522

READING CLAUSEWITZ

Beatrice Heuser is Professor of Strategic and
International Studies in the Department of War
Studies at King's College, London. She previously
taught at the universities of Oxford and Reims and
has also worked as a consultant to the Ministry of
Defence, the European Commission and NATO.

READING CLAUSEWITZ

BEATRICE HEUSER

PIMLICO

Published by Pimlico 2002

2 4 6 8 10 9 7 5 3 1

Copyright © Beatrice Heuser 2002

First published in Great Britain by
Pimlico 2002

Pimlico
Random House, 20 Vauxhall Bridge Road,
London SW1V 2SA

Random House Australia (Pty) Limited
20 Alfred Street, Milsons Point, Sydney,
New South Wales 2061, Australia

Random House New Zealand Limited
18 Poland Road, Glenfield,
Auckland 10, New Zealand

Random House (Pty) Limited
Endulini, 5A Jubilee Road, Parktown 2193, South Africa

The Random House Group Limited Reg. No. 954009
www.randomhouse.co.uk

A CIP catalogue record for this book
is available from the British Library

ISBN 0-7126-6484-X

Papers used by Random House are natural,
recyclable products made from wood grown in sustainable forests;
the manufacturing processes conform to the environmental
regulations of the country of origin

Typeset by Deltatype Ltd, Birkenhead, Merseyside

Printed and bound in Great Britain by
Mackays of Chatham PLC

For Eleonore

Contents

Foreword

[Clausewitz] makes it impossible, even for a weak head like mine that is otherwise no good for studying, not to follow him, and he directs you on leading strings from step to step, from conclusion to conclusion, so that you cannot go wrong or lose your way to the right or the left.

Sophie Countess Schwerin (1785–1863)[1]

The purpose of this book is to allow the reader to share the reaction of Sophie Countess Schwerin when she read Clausewitz's work. All too often, Clausewitz has been dismissed as terribly difficult to understand. As we shall see, his most famous work, *On War*, was undergoing major revisions at his own hand when in 1830 Clausewitz had to go on active military duty from which he was never to return. Left in the middle of this revision, the work contains important contradictions which have baffled many readers and which, worse still, have led to misinterpretations of enormous consequence.

It is my ambition in this book to distil from Clausewitz's writing what are arguably the most important ideas, to point out and explain the key contradictions, and to guide the reader through Clausewitz's writing. Moreover, I want to show what impact his ideas had on later generations of strategists throughout the world, some of whom distorted them beyond recognition, some of whom thoroughly misunderstood them, some of whom dismissed them, partly from ignorance, partly because their world-view dictated it, and some of whom, grafting onto their own thoughts ideas that were only buds on the stem of *On War*, allowed them to develop into green shoots, and occasionally into strong branches bearing rich fruit, which could, however, prove poisonous.

In the first chapter I shall acquaint the reader with Carl von Clausewitz, with what people said about him and about his most famous book. The second chapter will explain the most important ideas and contradictions of *On War*, the third with Clausewitz's ideas of state and society and the fourth will look at factors which Clausewitz discovered as having an important impact on war, such as the commander's military genius, or the morale of the forces. I will then grasp the thorny nettle of whether Clausewitz, who preached the superiority of defence over the offensive, can be blamed for the horrors of the First and Second World Wars, explore what other strategists have made of Clausewitz's ideas applied to naval warfare, to 'limited' wars of the twentieth century, and to irregular warfare, and take Clausewitz and us into the nuclear age. In the final chapter I shall deal with critics of Clausewitz's work, before assessing his importance for understanding war in the twenty-first century.

This book does not aim to add great original insights to our understanding of Clausewitz. Instead, by bringing together the gist of the most important existing publications, I hope it will furnish readers with a more complete overview of his thinking on war, its basic problems and its enduring merit. It is aimed primarily at students of strategy and war who want to understand the essence of what Clausewitz had to tell us, without having to read all the secondary literature. Anybody with a little more time on their hands is warmly encouraged to look at the notes at the end of this volume, which will indicate other literature on each respective subject.

I have chosen to translate the quoted passages from Clausewitz's writings afresh, so as to enable me more easily to stress the continuity in some of his themes than the freer and often more elegant translations in existence have done (and without doubt the most elegant existing translation into English of *On War* is that by the late Angus Malcolm, Peter Paret and Sir Michael Howard of 1976). The fact remains that Clausewitz was a lover of long and complex sentences in which he first expressed his important insights, and then added subclause upon subclause to accommodate exceptions and limitations to his rules. Also, as he himself invented or perfected a good number of concepts, we often find him groping for words and using many synonyms when referring to the same thing, when a thorough revision of the book might have led him to standardise his terminology. Thus he uses the words *Begriff, Idee, das Wahre* and other expressions for what I have translated as 'concept', and what others have called 'nature', 'essence' or 'ideal'. Clausewitz does not actually refer to 'absolute war' or 'limited war' but 'war in its absolute perfection' and 'limitations on the conduct of war', etc. He did not have Sir Michael Howard's admirable gift for reducing complex problems to concise, snappy and yet meaningful phrases. *On War* could have done with a good deal of editing to give it more focus and greater precision, which in her sorrow Clausewitz's widow and editor did not dare apply to the work. The result, oddly, is that the Malcolm–Paret–Howard translation is sometimes easier to read than the German original, lovingly restored to its early nineteenth-century vocabulary by the late Werner Hahlweg. It is that edition, the nineteenth (and unchanged twentieth) printed by Dümmler publishers (1980 and 1989 – respectively), which I have used as the basis of my translations.

I am very grateful to Will Sulkin and Jörg Hensgen at Pimlico for having encouraged me to write this book and for having put up with the delay in its completion. My colleagues Jan Willem Honig, James Gow, John Stone and Brian Holden Reid, who gallantly shouldered my teaching duties during my sick leave, indirectly allowed me to catch up with myself and deserve warmest thanks, as does Christopher Dandeker in backing me throughout this difficult time. My assistants, Michael Ploetz, Anastasia Filippidou and Andrew Monaghan, were very helpful and forbearing, and I am particularly grateful to the latter and to Andrew Lambert and Robert Foley for their comments and criticism. My parents and my husband were immensely supportive and tolerant. I owe thanks to the staff at the British Library, the University Library in Munich, and the Institute for Contemporary History in Munich, and to Mr Mecklein of the local library in Rothenburg ob der Tauber, who knows more about inter-library loans now than he ever wanted to. And finally, I am grateful to Eleonore for learning, not a day too soon, that books are not for tearing up but for reading and treasuring, which is how my own copy of *On War* survived her fascinated attention.

1

The Story of the Man and the Book

> Should an early death interrupt me in this work, then [the manuscript]
> that will be found will admittedly only be a formless mass of ideas,
> which, subject to endless misunderstandings, will lead to a great deal of
> premature criticism . . .
> But despite the incomplete form I believe that an unprejudiced reader . . .
> might find in it . . . ideas which might engender a revolution in the
> theory [of war].
>
> (Clausewitz in a note of 10 July 1827)

Clausewitz's Early Life

Carl Philipp Gottlieb[1] von Clausewitz was born into a large family on 1 June 1780 in the small Saxon town of Burg, near Prussian-held Magdeburg. On his father's side, the family descended from minor nobility that originally came from Upper Silesia, but their claim to nobility may have been contested, as Carl's paternal grandfather, who was professor of theology in Halle, made no use of the particle 'von' in his name. Carl's father, however, wanted to revive his claim to the title, and petitioned Frederick the Great of Prussia to allow him to use the 'von' again. The king agreed and employed Carl's father in one of his regiments, where he made a very modest career. He 'was an officer in the Seven Years War, and he was full of the prejudices of his profession.' In his parents' house, young Carl 'saw almost nobody but officers (and not exactly the most educated . . . ones).' As Carl himself wrote, he had 'grown up in the Prussian army.' At the age of twelve, Carl 'himself became a soldier, [and] took part in the campaigns of 1793 and 1794 against France.' Carl thus came to see 'war in his youth, without understanding it, yet an impression of the whole . . . remained with him.'[2]

1

Besides Carl, two of his three brothers also became army officers, like Carl rising to the rank of general. At the age of twenty-one Carl was transferred to the War College (*Kriegsakademie*) in Berlin, which was directed by General Gerhard Johann David Scharnhorst (1755–1813), who became his mentor and teacher, drawing his attention to the link between the revolution in French society and the new French way of war. Scharnhorst campaigned for a thorough reform of the Prussian military system, and, like a number of his disciples, Clausewitz soon shared this reforming spirit, with which they antagonised their superiors and the crown.

The first and main enemy Clausewitz encountered in battle was France, and from his earliest youth he was thus at once full of dread of the French way of war and full of hatred for the French, who under Napoleon set out to conquer Europe. Even in 1803 he wrote a combative piece in which he compared the French with the tyrannical and imperialist Romans of old.[3] Writing in 1808, he well summed up what was increasingly the German reaction to the Napoleonic occupation, once the original admiration for the Corsican reformer had died down: 'There has never been a nation which reacts differently than with hatred and enmity to the direct pressure on it coming from another nation.'[4] Clausewitz used the whole panoply of contemporary stereotypes in his denigration of the French: he described them as shallow, uniform, frugal, vain, easy to force into obedience to the wishes of the government and thus good political instruments. The Germans by contrast he described as individualistic, original, keen on reasoning and arguments, hard working, endlessly striving for higher goals. The French he compared with the practical Romans and the Germans with the more intellectual and morally inspired Greeks.[5] Even shortly before his death he was still writing plans for war against France:[6] at the time when he was trying to finish *On War*, the 1830 Revolution in France led Clausewitz to fear that 'in France new revolutions will develop from the loins of the first revolution, and that this country will then break the peace.'[7]

Much of his Francophobia he owed to his wife, Marie Countess von Brühl, whose English mother 'was stimulated by all the hatred of Napoleon that characterised English people at the time. She and her daughter, Marie Brühl, later Frau von Clausewitz, lived entirely in the fervour of politics and the passionate hatred of the French', as the Berlin socialite Caroline von Rochow noted.[8]

Clausewitz met Marie in 1803 when he was twenty-three. Although they fell in love with each other almost immediately, they had to wait many years before they could get married because of the differences in social standing between them. As we have noted, Clausewitz was of very minor, even dubious nobility, which was only confirmed for him by King

Frederick William III in 1827. The Brühls, by contrast, were among the very highest nobility, which for seven years proved an obstacle to Clausewitz's love match with their daughter. Both during the time of courtship and later in married life, the well-read Marie had considerable influence on Clausewitz's thinking on a host of matters, and, as the granddaughter on her mother's side of a British consul to Russia, this included a great admiration for Britain, to be matched only by Clausewitz's hatred of all things French. Carl and Marie finally got married in December 1810, and remained great soul-mates until the end of his life.

On 14 October 1806, Clausewitz took part in the battle of Auerstedt, his first experience of the Napoleonic style of battle. Prussia was defeated, Napoleon entered Berlin and concluded a pact with the neighbouring kingdom of Saxony, before moving against Russia. As companion of the Prussian prince Augustus, Clausewitz was taken prisoner after the battle and had to go to France, where he was allowed a comfortable life as a gentleman-prisoner of war. Nevertheless, this experience only confirmed him in his anti-French sentiments, as his letters to his fiancée demonstrate.[9]

The battle of Eylau in February 1807 being inconclusive, Napoleon signed a peace treaty with Russia, dividing Europe (and the Hohenzollerns' dynastic possessions) into two spheres of influence. As a consequence of this settlement, Prince Augustus and Clausewitz were repatriated in 1808, and Clausewitz rejoined his mentor Scharnhorst.

In 1810–11 Clausewitz was in charge of lectures on 'small war' (the term would become more famous as *guerrilla*, as the Spanish called their resistance to Napoleon) at the general school of war (*Allgemeine Kriegsschule*) in Berlin. Between 1810 and 1812 he tutored the Prussian crown prince, the future King Frederick William IV, and in this context applied his thinking to many of the issues surrounding war which would later form the basis of his famous work. The curriculum for the crown prince was fairly classical: Clausewitz lectured his tutee about where one should deploy artillery and how, about soldiers' motivation, about defence and attack. Here Clausewitz was beginning to theorise about war rather than just about battle, as most of his predecessors had done. His intellectual star was rising.

But his anti-French convictions became a stumbling block to a more successful career in Prussia: in April 1812 he felt so outraged by the kowtowing policies towards France of King Frederick William III that he resigned from the Prussian army, and wrote a fiery political 'Confession' which incorporated all the sentiments of *Sturm und Drang* and of the budding Prusso-German nationalism. 'I believe and profess', he wrote,

that a people should respect nothing more than its dignity and its freedom; that it should defend these with its last drop of blood; that it has no holier task to fulfil, no higher law to obey; that the shame of cowardly submission can never be expunged; . . . that one can only lose one's honour once; that the honour of the king and the government are one with the honour of the people . . . [10]

As Napoleon was preparing for a new campaign against Russia, Clausewitz offered his services to Tsar Alexander I, even though he did not speak any Russian, and was accepted as an officer into the imperial army. On 7 September 1812 he witnessed what was perhaps the bloodiest of all of Napoleon's battles, the battle of Borodino, as a result of which he later wrote an insightful eye-witness account of Napoleon's entire Russian campaign. In December of that year, Clausewitz made his only ever excursion into the world of diplomacy, when he took part as one of the chief negotiators under General Hans David Ludwig von Yorck in the peace negotiations between Russia and Prussia at Tauroggen, leading to the anti-French alliance which was the basis of the Wars of Liberation (1813–14). As an imperial officer he took part, under the command of Blücher, in the battles of Groß-Görsch (2 May 1813) and Bautzen (20–21 May 1813). His experience in 1812–13 of the bitter Russian winter left him with a ruddy complexion and red nose, to the extent that uncharitable spirits spread the rumour that he began each day with a bottle of wine.[11]

Writing On War

Even though Napoleon was defeated in 1814, Frederick William III was suspicious of those officers who had left Prussia for service in other armies before Prussia openly rose against France. Thus Clausewitz was only admitted back into the Prussian general staff as a colonel in 1815. Subsequently rising to the rank of general, he was appointed director of the *Allgemeine Kriegsschule*, where he was charged with much administration, and a little teaching, research and writing. Clausewitz took this as a slight: he would have much preferred an active military post. But it did give him time to write *On War*, which he began to write in 1818 and last added to in 1830. As we have seen, before having revised it to his own satisfaction, in 1830 he was called to active duty to counter a Polish uprising against the Russian occupiers which threatened to spill over into Prussian-occupied lands. Clausewitz reached Posen, then still part of East Prussia, now as Poznan part of Poland, after the uprising had been quashed, but stayed there and in Breslau (now Wrocław) until the following year, along with his second mentor, A. Neithardt von Gneisenau. Both he and Gneisenau fell victim to a cholera epidemic, Clausewitz dying in Breslau very suddenly on 16 November 1831. His unfinished manuscript was subsequently published by his widow in 1832–4, without any alterations, as Parts I–III of his *Posthumous Works about War and the Conduct of War*.[12] In a note of 10 July 1827, Clausewitz had written prophetically of this fear that if 'an early death' should prevent the completion of the manuscript, it would lead to 'endless misunderstanding' and 'immature criticism'.

> For in these matters everybody believes that what has just come into his mind . . . is good enough to be said or printed, and thinks it just as evident as the fact that two plus two makes four. If they took the trouble like me to spend years thinking about the subject and to compare it time and again with the history of war, they would be less rash in their criticism.

He nevertheless hoped that even in incomplete form, the manuscript already contained the theses which 'might engender a revolution in the theory of war'.[13]

Contemporaries held quite ambivalent views of Clausewitz. One Wilhelm von Dorow wrote to a friend: 'Clausewitz, as you know him also, is a deeply scholarly, educated, pure military person of note, with a most endearing, conciliatory, decent character.' By contrast General von Valentin found 'this man . . . totally unpalatable'. Caroline von Rochow, *née* von der Marwitz, shed some light on how this military scholar could impress some as very sophisticated and polished, and yet be uncongenial to several of his military colleagues. Clausewitz, she wrote,

> had quite a disadvantageous personality and his exterior was cold and unengaging . . . He spoke little, conveying the impression that his interlocutors and the subjects were not good enough for him . . . At the same time he was filled by burning ambition and yearned more for classical forms of self-fulfilment than for the modern fashion of stimulating enjoyment. He had few, but close and reliable friends, who had higher hopes and expectations for him than . . . fate or circumstances or his unprepossessing personality could achieve for him.[14]

Countess Bernstorff found Clausewitz a social liability when she entertained, as his 'likes and dislikes, [expressed] in their painful decisiveness, more or less always influence the spirit of the party' and had over time 'occasioned several embarrassing incidents'. He tended to treat others in a very abrupt manner or indeed with great reserve, yet was extremely touchy if others treated him in the same way, and quick to take offence. The

Prussian officer von Scharfenort, who knew him at the Prussian General War School and was on the whole well disposed towards him, commented on his 'often caustic sense of humour', on his sarcasm 'which could hardly be surpassed in sharpness and which can make him many enemies'. When, in 1818, Clausewitz tried to become Prussian minister in London, Frederick William III had little inclination to send somebody who had so often been in opposition to his policies. J. A. Rose, the British minister in Berlin, in his letters home commented thus on Clausewitz's 'cold and distant habits': 'His manner is cold and by no means popular; his temper is said to be vehement.' Finally the Duke of Cumberland, brother of the British King George IV, noted that Clausewitz could hardly occupy 'such a high position as Minister at the Court of St James, for he is not made for society.'[15] One scholar points out that Clausewitz (after the temporary setback he suffered because of his self-imposed exile of 1812–15) became a major general at the age of forty, and yet he constantly complained to his wife about being slighted, ignored, even though on the several occasions when other officers were preferred to him, they tended to be older, more experienced men.[16] Clausewitz was thus a difficult character, more at ease in a library than in Berlin's vibrant salons or among most of his fellow-officers.

On whose works did Clausewitz construct the edifice of his own thinking? Clausewitz's education at the War College under Scharnhorst was comprehensive, as is reflected in his writings. The American Clausewitz specialist Peter Paret thought that 'Clausewitz was a typical educated representative of his generation, who attended lectures on logic and ethics designed for the general public, read relevant non-professional books and articles, and drew scraps of ideas at second and third hand from his cultural environment.'[17] Thus we find references to strategists like Montecuccoli, de Saxe, Feuquières, Santa Cruz, Puységur, Turpin, Guibert, Frederick the Great, Prince de Ligne, Maurillon, Venturini, de Silva, Lloyd and Berenhorst, the *doyen* of the new wave of Prussian military writers.[18] Among the political philosophers whose works Clausewitz read were Robertson, Johannes von Müller, Ancillon, Gentz, Montaigne and Montesquieu (whose *De l'Esprit des Lois* he consciously took as his model for *On War*[19]). He also read Kant, although he might not have been as greatly influenced by him as some would argue.[20] The relationship between Clausewitz and Hegel has been much debated, the German Clausewitz specialist Schering pointing out commonalities in their writings.[21] As we shall see, the Soviets were also convinced that Clausewitz was influenced by him. And yet there is no hard evidence that Clausewitz read Hegel or heard him lecture, nor are his ideas on the 'dual nature of war' in any way to be understood as dialectic or his ideas on absolute war inspired by Hegel's 'absolute ideas' (for which Lenin may have mistaken them).[22]

By contrast, we know for certain not only that Clausewitz had read Machiavelli, but that he was particularly taken with him, as he entered into correspondence about him with the philosopher Fichte, who had published an article in defence of Machiavelli.[23] Clausewitz had read *The Prince*, the *Discorsi* and his *Arte della Guerra*. He shared Machiavelli's amorality in his discussion of war – famously, Machiavelli in his *History of Florence* had called 'those wars just that are necessary, and those weapons holy that are our last hope' (Vol.8). Clausewitz wrote:

> No book on earth is more necessary to the politician than Machiavelli's. Those who affect disgust for his principles are a kind of minor spirits of humanism. What he says about the princes' policies towards their subjects is certainly largely outdated because political forms have changed considerably since his day. Nevertheless, he formulates some remarkable rules, which will remain valid for ever . . . But [Machiavelli] is especially instructive with regard to foreign affairs, and all the scorn cast upon him (to be sure, often out of confusion, ignorance, affectation) concerns his teaching on the treatment of the citizenry, insofar as it is justified at all.

To give Machiavelli the credit due to him, one had to take into account the circumstances of Renaissance Italy:

> If we have regard for all these conditions, we will not be able to accuse him of more than that, with a certain lack of prudishness, he called things by their proper name . . . The 21st chapter of Machiavelli's *Prince* is the basic code for all diplomacy – and woe to those who fail to keep to it![24]

Chapter 21 of *The Prince* is headed 'How a Prince must act to win honour' and begins: 'Nothing brings the prince more prestige than great campaigns and striking demonstrations of his personal ability.' Most importantly, in this chapter the great Florentine advised strongly in favour of taking sides in the conflicts of others and of avoiding neutrality,[25] which seemed most prophetic to Clausewitz in the circumstances of his own prince's decision to come to an arrangement with Napoleon, rather than to forge or join a coalition against him.

Did this make Clausewitz a cynic in his analysis of politics and inter-state relations? Not if we see him against the background of political thinking in his time. His life straddled two very different periods, two distinct mentalities. He was born into the world of the *ancien régime*, in which soldiers, artists and other professionals sought their fortune at any court of Europe that would have them, in which the *lingua franca* and the currency of political thought was French, and in which people owed their loyalty to their princes, their guild, their town. It was this cosmopolitanism among the professionals which allowed men like the Prussian Clausewitz and the Swiss Jomini to seek employment in the Tsar's forces. Yet this very move on the part of Clausewitz, albeit smacking of the cosmopolitanism of the *ancien régime*, was made in the spirit of a new age: that in which nationalism replaced regnalism (loyalty to one's prince). For Clausewitz the monarch was the mere representative of the nation, the incarnation of its honour and glory, but no longer the representative of God on earth as he had been thought of in previous times.[26] Thus Clausewitz's departure for Russia in 1812 was symbolic of his devotion to the cause of German liberty from French oppression, which led him to leave his king to serve another who in his view was doing more to withstand his own enemy and that of Prussia.

We have already seen that Clausewitz abhorred the egalitarian ideals of the French Revolution. But for his time he was no extreme reactionary either, and thought that an oligarchic system would be the best for Prussia. 'Let the government gather around it representatives of the people, elected from those who share the true interests of government and are known to the people. Let this be the government's principal support, friend and ally, as Parliament has been for a century the support of the king of England.'[27] (Again we see the influence of his half-English wife.)

Clausewitz was very much in line with the new philosophical school of Fichte and Hegel in arguing that 'the State's honour must be regarded as holy' (which the National Socialists later quoted with reverence).[28] Peter Paret called him 'a rabid patriot' who came 'to see the state as a pure organism of power, bare of all ideological decorations'.[29]

Writing in the early 1820s, Clausewitz was already desperately awaiting the unification of the many little states in Central Europe in which German was spoken: 'Germany can reach political unity in *one* way only, through the sword, when one of its states will become master of all the others.' But Clausewitz thought the time for this had not yet come, nor was he prepared to predict which of the German states would dominate the others.[30]

How is Clausewitz different from previous strategists? Why do we still give him so much attention compared with other thinkers (Machiavelli, Guibert, all the eighteenth-century 'tacticians', and his contemporary Henri Baron de Jomini (1779–1869), whose writings were much more popular in the first half of the nineteenth century)? Like most outstanding works, Clausewitz's *On War* was the best of a batch of books on the subject. These included works by Adam Heinrich Dietrich von Bülow (1757–1807), another Prussian. Bülow was most explicitly interested in mathematical rules and geometrical formations. He is chiefly remembered for having claimed, in his *Spirit of the Modern System of War* (1799, first English translation 1806), that 'it is always possible to avoid battle'.[31] In an essay of 1805, Clausewitz criticised Bülow severely on a variety of points. Bülow had defined strategy as 'the science of martial movements outside the field of vision of the enemy, tactics within [his field of vision]'; Clausewitz regarded these definitions as too mechanical. He dismissed Bülow's belief that battles did not decide matters and could be avoided as plain wrong, his view that it was the pinnacle of the art of war to achieve 'great effects through small means' as too detached from the balance of forces in any given situation, and his lack of interest in morale and his emphasis on mathematical calculations as immature. Rather cruelly, Clausewitz wrote that Bülow might be called 'The children's military companion'.[32]

Another contemporary was Archduke Charles of Austria (1771–1847), who had been the most successful general in the fight against Napoleon, achieving victory for Austria in the battle of Aspern Essling, and was much admired. Besides his own considerable military achievements on the battlefield, he contributed to the literature on warfare by publishing *Principles of the Higher Art of War* in 1806, and *Principles of Strategy* in 1814. He worked with many of Bülow's ideas, particularly with his geometrical discussion of operations.[33]

More works on war were produced by the Prussian military reformers who had lived through the French Revolutionary and Napoleonic Wars, had seen the ignominious and pitiful defeat of their own country, and had concentrated their energies on the reform of the Prussian military to make it fit to fight for the liberty of their country from French occupation. While Georg Heinrich von Berenhorst (1733–1814) is usually named as the first of these, Scharnhorst was the teacher of several of them, including Clausewitz himself and Otto August Rühle von Lilienstern (1780–1847). Berenhorst was the illegitimate son of Prince Leopold I of Anhalt-Dessau. He took part in the Seven Years War, and later wrote *Reflections on the Art of War* in three volumes (1796 and 1798², 1798, 1799). Like Clausewitz after him, he stressed the importance of the morale of troops, of the decisiveness of the personality of the leader, and of chance or accidents.

Scharnhorst also alerted his students to the psychological and political factors in war, to the importance of the morale of the armed forces and the genius of the commander. He, too, was prolific, and his works included a handbook for officers. In 1797, together with Friedrich von der Decker, he wrote a 'Development of the General Reasons for the French Success in the Wars of Revolution', in which he analysed in some detail the appeal of the French Revolution and its ideals to intellectuals, the young, and those who showed concern for social equality, explaining how the ideals served to inspire them in their political activities but also in their warfare. From 1801, he interested his disciples at the Kriegsakedemie in Montesquieu's *Spirit of the Laws*.[34] Furthermore, Scharnhorst believed that there were particular principles that could be distilled from the data of previous wars: 'The great generals of history – Hannibal, Scipio, Caesar, Turenne, Montecuccoli and Frederick – studied the principles of the art of war.'[35]

Another Prussian writer of this period was General Friedrich Constantin von Lossau (1767–1833), who in 1815 published *Der Krieg* ('War'). Azar Gat summed up this book as dealing 'extensively with the warrior's intellectual and moral faculties, presenting war as a clash of wills motivated by patriotic and other psychological energies.'[36]

As we shall see in the following chapters, many of Clausewitz's key ideas were built on the insights of those mentioned above. Yet he achieved a substantially greater level of abstraction than most of his peers by deriving his conclusions about war from the evidence of 130 historical battles which he had studied closely, and yet without allowing himself to become bogged down in detail. Unlike the other writers of his generation, Clausewitz put most emphasis on the changing nature of strategy and tactics throughout history. When he mentioned Alexander, Caesar, Frederick the Great and Napoleon in one breath, it was only to reflect on their military genius, or the importance of discipline in their armies, rarely to attempt to prove that they had all applied the same principles to lead them to success (such as 'inner lines' or 'the indirect approach'); indeed, Clausewitz explicitly dismissed any monocausal explanations of their success.[37]

This is how Clausewitz formulated his criticism of many thinkers of the previous generation:

> Previously, the terms art of war or sciences of war only denoted the sum of that knowledge and those skills which concern material things. The design and production and the use of weapons, the construction of fortresses and defensive positions, the organisation of the army and its movements were the subject of this knowledge and these skills . . . All this is in relation to fighting not much different from the art of the sword cleaner to the art of fencing . . . one did not speak of the actual movements of the spirit and of courage.[38]

Clausewitz's contemporaries and both earlier and later strategists, it is true, had a tendency to set down rigid rules of behaviour, which on the whole would have proved too narrow to withstand the changes in warfare brought about by innovation in technology, for example. The pointless charges of the First World War are evidence of the rigid and unintelligent applications of such rules, rather than of an intelligent reconsideration of basic premises, as Clausewitz would have advised. Clausewitz himself aimed to produce a theory of war that was to be a contemplation of the subject, not a doctrine according to which one should act in specific cases: the theory should produce a familiarity with the subject which in turn should turn into *Können*, the ability to do, and *Kunst*, art, without becoming a constraining corset of rules and regulations.[39]

In his *Précis de l'art de la guerre*, published in 1838 (thus after Clausewitz's *On War*), Jomini (Clausewitz's main Francophone competitor), who had fought under Napoleon and, like Clausewitz, in the imperial Russian army, wrote scathingly that Clausewitz 'in fact shows himself as too sceptical of military science: his first volume is nothing but a declamation against any theory of war, while the two following volumes, full of theoretical maxims, prove that the author believes in the efficacy of his doctrines, even if he does not believe in that of others.'[40]

And indeed, just as Jomini alleged, Clausewitz did formulate a number of tenets which are reflected in his *opus magnum*. The most concise summary of his propositions is found in a note Clausewitz wrote, probably in 1827:[41]

- that defence is the stronger form [of warfare] with the negative purpose, attack the weaker form with the positive purpose;
- that the great successes also determine the smaller ones;
- that one can thus trace back strategic effects to certain centres of gravity;
- that a demonstration is a weaker use of force than a real attack, that it must therefore be based on particular conditions;
- that victory does not merely consist of the conquest of the battlefield but of the destruction of the physical and moral fighting forces, and that this is mostly only achieved in the follow-up to the battle that has been won;
- that success is always the greatest where the victory was achieved by fighting;
- that therefore the change from one line and direction [of attack] to another can only be regarded as a necessary evil;
- that an attack on the rear [*Umgehung*] is justified only by [one's own] superiority in general or by the superiority of one's own line of communications and retreat over that of the adversary;
- that flanking positions are conditional upon the same circumstances [as spelled out in the previous point];
- that every attack weakens as it progresses.[42]

With these propositions (*Sätze*), Clausewitz came close to establishing his own simplistic rules. Moreover, what he described as the essence of war in *On War* lent itself to the deduction of rigid, simplistic rules, as we shall see.

Nevertheless, more than any other strategist of the period, Clausewitz wanted to get away from simple, mechanistic rules. His purpose was to teach his reader how to think about war, rather than give prescriptions for set battle pieces. Particularly in his lessons for the future King Frederick William IV of Prussia, Clausewitz wanted to teach his tutee what to keep in mind when he would have to decide for himself what to do, not give him set answers.

Clausewitz stated himself that his purpose in writing *On War* was 'to iron out many a crease in the heads of strategists and statesmen' where their understanding of war was concerned; it was to 'educate the mind of the future military leader or rather give him guidance in his self-education, but not accompany him onto the battlefield, just as a teacher guides and facilitates the spiritual development of a youth, without, however, keeping him strapped in leading strings all his life.'[43] In many respects, Clausewitz is thus providing something that few military practitioners are on the lookout for. While they tend to be in search of teachable and learnable rules of thumb that can be applied to a wide range of different situations, and can help them find short-cuts to decision-making in stressful combat situations, Clausewitz mainly supplies philosophical reflections on the nature of war that are difficult to translate into simple, memorable prescriptions for action. As a result, *On War* met with far less appreciation in the first century after its publication than one would think, judging by the high esteem in which Clausewitz has since been held, particularly in the late twentieth century.

The Reception of On War

When *On War* was first published in 1832–4, it caused a stir in the German-speaking world. Right from the beginning, even his admirers were critical, admitting that the 'nuggets' in Clausewitz's writing were sometimes difficult to find in the gush of prose, and that it hardly lent itself to superficial study. An early reviewer thought that 'only the determined scholar' could hope to reach an adequate understanding of the work. A contemporary military review commented, 'The language used by the author is very sophisticated, but not popular enough to be understood by the great many. This work thus not only wants to be read, it wants to be studied.' Sophie Countess Schwerin (quoted in the Foreword), was in a small minority when she wrote that Clausewitz made it easy for his readers to follow his logic. Writing in 1838, the French military author J.

Rocquancourt noted, 'The Prussians who have made a lot of noise about [*On War*] are beginning to recognise all its shortcomings.'[44] In the same year, Jomini wrote grudgingly,

> One cannot deny to General Clausewitz great learning and a facile pen; but this pen, at times a little vagrant, is above all too pretentious for a didactic discussion, the simplicity and clearness of which ought to be its first merit. Besides that, the author shows himself by far too sceptical in point of military science . . .
>
> As for myself, I own that I have been able to find in this learned labyrinth but a small number of luminous ideas and remarkable articles; and far from having shared the scepticism of the author, no work would have contributed more than his to make me feel the necessity and utility of good theories.[45]

On War secured its place in Prussian military education under the influence of Helmuth Graf von Moltke (known as 'the elder Moltke' to distinguish him from his homonymous nephew). When the elder Moltke became Prussian chief-of-staff, he was very concerned not only that the Prussian military staff have a good education, but that *On War* be added to the curriculum. He told a French journalist that Clausewitz had made the greatest impact on him after the Bible, Homer, Littrow and Liebig.[46] Consequently, with *On War* a standard text in the Prussian officers' education, the Prussia military victories of 1864, 1866 and 1870–71 were variously attributed to the influence of Clausewitz.[47]

But it was not merely military officers who read *On War*. Karl Marx and Friedrich Engels both read Clausewitz, and grudgingly came to appreciate him. Engels wrote cryptically: 'Bizarre way of philosophising, but very good treatment of the object'.[48] He thought Clausewitz's writing was 'of a very high standard', and counted him among 'the classics in his subject area along with Jomini.' Elsewhere he called him 'a first rate star'.[49] Marx in turn thought Clausewitz had 'common sense . . . bordering on wit'.[50] Nevertheless, Engels referred far more frequently to Jomini than to Clausewitz, and wrote in a personal letter in 1853, 'I can't really get myself to like that natural genius, Clausewitz.'[51] Engels' writing also shows that he was not particularly influenced by the key ideas of Clausewitz's work. Thus he wrote in 1855, after reading *On War*, 'The [military] chief who is influenced by political considerations or who acts with a want of resolution must lose his army.'[52] (Engels, incidentally, was very much aware of the effects, both physical and social, which changes in military equipment could bring about, and devoted particular attention to the revolutionary effects of fire power,[53] where Clausewitz paid little heed to the effects of such technical changes on war.) Following in the footsteps of Marx and Engels, Lenin admired Clausewitz greatly, as we shall see.

In the French-speaking world, it took much longer for Clausewitz to become popular, not only because of the significant language barrier – writing in 1840, Rocquancourt dismissed Clausewitz's writing as 'at times a little unintelligible'[54] – but also because he had well-placed competition from Jomini, who was the main Francophone analyst of Napoleon's way of war. Apart from his language, Jomini had longevity on his side. He was thus infinitely more influential than Clausewitz, particularly outside Prussia, until the late nineteenth century,[55] and remained the most frequently read interpreter of Napoleon's strategy in France and in the United States.

And just as Clausewitz disliked French for the way in which the grammar imposed patterns of thought alien to him, Francophones found his German prose impenetrable. The oldest son of France's King Louis-Philippe, the Duke of Orleans, commissioned a translation of *On War* which was never completed, but also a summary, as *Résumé des principes de Clausewitz*, significantly written in French by a Polish emigrant, Louis de Szafraniec Bystrzonowski, which was published in the periodical *Le Spectateur militaire* over several months in 1845, and then in book form in 1846. The first full translation was made in 1849–51 by a Belgian, the commandant Neuens. The first extensive commentary, this time at last by a Frenchman, Captain de la Barre-Duparcq, followed in 1853, but his appreciation was very critical. He described *On War* as a metaphysical treatise bringing together an immense number of military considerations with little to link them, presented as based on historical examples, but without explicit demonstrations. As a result he thought that 'the enlightened study of [*On War*] would be profitable for [experienced] officers . . . to enlarge the circle of their theoretical ideas on war; but where young officers are concerned, I do not advise them to read [*On War*] because it could throw their minds into confusion.'[56]

Despite this translation, Clausewitz was little read in France until the 1880s. After the debacle of the Franco-Prussian War of 1870–71, French attention was mainly focused on copying or in some ways learning from the Prussian armed forces.[57] Many reforms were made in the structure and administration of the French army, including the creation of a new general staff, and the brigade-division-corps structure which Prussia had developed. The staff college was reformed in 1876–8 and renamed *École supérieure de guerre* in 1880, modelled on the Prussian *Kriegsakademie*. General Jules Lewis Lewal, the founding director of the *École de guerre* in 1877 and minister of war for a short period, assumed Clausewitz into the pantheon of great military thinkers – with de Saxe, Guibert, Jomini and others. Of the Clausewitzian concepts, Lewal agreed with Clausewitz's emphasis on the importance of battle, on the concentration of forces, and on flexible, offensive defence, but was most enthusiastic about the role of the will, *la volonté*, in the French sense, in which will-power mattered more than numbers.[58] But in their soul-searching after the defeat of 1870–71, most French strategists, including Adolphe Thiers, historian and first President of the French Third Republic founded in 1871, soon went back to Napoleon and to his other interpreter, Jomini, or else studied Napoleon's strategies afresh on the basis of historical evidence. Thus General Berthaut, reflecting on the *Principles of Strategy* (the title of his work of 1881), did not mention Clausewitz once, although he borrowed amply from other military authors, particularly Napoleon, Archduke Charles and Jomini.[59]

The *École de guerre* was more vocational than the German equivalent, did not encourage the students to think independently, and did not teach a particularly coherent doctrine.[60] Nevertheless, Clausewitz's influence spread through this institution. Staff who took an interest in Clausewitz, apart from Lewal, included Cardot, Maillard, Bonnal, Langlois, Cherfils and Niox. The next generation, that of their pupils, included Foch, Leblond, Lanrezac and Ruffey.[61] Major Lucien Cardot lectured on Clausewitz there in 1885. But the majority of French strategists remained sceptical of Clausewitz. Henri Bonnal, lecturer, later the school's commandant, and contributor to the French war plans for the First World War, criticised Clausewitz for not giving enough attention to Napoleon's enveloping strategy, the *manoeuvre sur les derrières*, and for a somewhat crude description of Napoleon as always going directly, rather than on occasion indirectly, after the enemy's forces. He thought the Prussian army was equally crude in its strategy, that Napoleon had been more subtle, and that it would be useful for France to follow the Napoleonic example, rather than the Prussian/German one.[62]

The second French translation of *On War* was only published in Paris in 1886–7, as *Théorie de la grande guerre* by Lieutenant Colonel de Vatry. As he wrote in his introduction, 'It is the painful memories of our disasters [the war of 1870–71] and the search for the causes which have brought them about which has inspired the idea to popularise a work from which the victor of 1871 has taken more than one lesson.'[63] The French military author Guillon explained: 'It is at the school of Clausewitz that the victors of Sadowa and of the Franco-German war were educated. Our generals had stayed behind at that of Jomini' while Clausewitz of course had been a 'direct teacher of Moltke'. In the following years Clausewitz enjoyed a little more popularity among the French military elite, to the point where Guillon, in a book entitled *Our military writers* published in 1899, called Clausewitz the greatest military writer of the century after Napoleon.[64]

Along with Clausewitz, Wilhelm von Blume and in particular Colmar von der Goltz were read widely among French military experts. (German was the only foreign language that had to be learnt at the *École de guerre*.) But the spell of French enthusiasm for Clausewitz did not last long, and remained an elite phenomenon in France.[65] Even in 1911, Colonel Camon wrote 'Today, he is no longer read.'[66] Many other leading strategists showed at most an indirect acquaintance with a handful of quotations from Clausewitz, taken out of context. For example, Admiral Raoul Castex of the French Navy, in his seminal book *Strategic Theories*, quotes Clausewitz only 10 times in over 400 pages, including the wonderful line: '"War is a violent form of policy" or "war is nothing other than policy continued in arms" (it is, I believe, Clausewitz, who said this or something similar).' (As his translator, Eugenia Kiesling, noted, Castex even misspelt Clausewitz's name[67] – although it has to be said that the misspelling of foreign names is widespread in France.) This would only change well after the Second World War, when it became chic for French intellectuals to have a view on Clausewitz.

In the Anglophone world, the first public figure to champion Clausewitz was John Mitchell (1785–1859), later a major general in the British Army, who had not only read some of Clausewitz's work – mainly on military history, it would seem – but described it, in 1839, as exemplary.[68] Mitchell commented that Clausewitz had written 'a very able, though lengthy, and often obscure book on War', and took on board Clausewitz's characterisation of war as a social activity rather than an art or science.[69] Otherwise the British staff college was, through works of Sir William Napier, Sir Patrick MacDougall and Sir Edward Bruce Hamley (all military officers), influenced by the Jominian tradition until the end of the nineteenth century.

On War was first published in Britain in 1873, translated by Colonel James John Graham, but sold only a few copies.[70] There is little evidence that he was widely read before the turn of the century. In 1889, Lieutenant-Colonel G. F. R. Henderson, lecturer in military history at Camberley, criticised this tradition, introducing major works from the German school.[71] In a lecture in 1894 he called Clausewitz 'the most profound of all writers on war', but added sarcastically, 'Clausewitz was a genius, and geniuses and clever men have a distressing habit of assuming that everyone understands what is perfectly clear to themselves.'[72]

In the following years Clausewitz was popularised in Britain indirectly through the interest shown in his writings by Chichele Professor of the History of War at Oxford, Spencer Wilkinson. There was a burst of interest in him just after the Boer War, peaking in the decade or so before the Great War,[73] and, as we shall see in Chapter 6, he became particularly important in influencing the thinking of Sir Julian Corbett, Britain's most important and interesting writer on maritime strategy at the time. After the

First World War, however, Clausewitz became a bogeyman owing to the very widely read Captain Sir Basil Liddell Hart, who blamed Clausewitz for the wastefulness of strategy during that war. As the American Clausewitz specialist Christopher Bassford has argued, Britain developed a 'tradition of Clausewitz-bashing'. J. F. C. Fuller, next to Liddell Hart the most famous British strategic writer of the inter-war period, called *On War* 'little more than a mass of notes, a cloud of flame and smoke'.[74] Another world war and several smaller wars later, Fuller relented, writing to a friend that in his opinion, 'Clausewitz's level is that of Copernicus, Newton, Darwin – all were cosmic geniuses who upset the world.'[75]

Most recently, the popular military writer John Keegan took up this baton of 'Clausewitz-bashing'.[76] But Britain is also the home of some of Clausewitz's greatest admirers. These included Field Marshal Claude Auchinleck (1884–1981),[77] Professor Sir Michael Howard, who founded the Department of War Studies at King's College, University of London and was one of Wilkinson's successors at Oxford, and Professor Colin Gray, who advised the American Reagan administration on strategy.

The story of the reception of Clausewitz in the USA is similar. The Jominian tradition went back to 1817, when a summary of Jomini's work was first published in America. At West Point military academy, Jomini became the chief text from which Napoleonic strategy was taught. During the American Civil War, military leaders on both sides were obsessed with the quest for a decisive battle in the Jominian/Napoleonic style, from Major General Henry Wager Halleck on the Union side to General Robert E. Lee on the Confederate side.[78]

In 1873, Colonel J. J. Graham's English translation of *On War* was distributed in the USA, although it was only in 1943 that an indigenous American translation by O. J. Matthijs Jolles was published. Lieutenant Colonel Robinson wrote as early as 1928 that 'a little research, a little study and reflection brings out the fact that Clausewitz's book on war, published in 1832, occupies about the same relation to the study of the military profession as does the Bible to all religious studies.'[79] But it was only in the late 1950s and early 1960s that American strategists widely began to base their reflections on Clausewitz's tenets. One of them, Bernard Brodie, wrote,

> Clausewitz's appeal is limited, for he is much more given than Jomini to 'undogmatic elasticity' in his opinions, and he is more metaphysical in his approach. Although an active professional soldier, he wrote with competence on philosophical problems pertaining to the theory of knowledge and in his military writings he sometimes used the technical idiom of the professional philosopher of his time. Besides, his insights,

like those of all great thinkers, can be fully appreciated only by readers who have already reflected independently on the same problems ... [Clausewitz dwells] as tellingly on the qualifications and exceptions to the basic ideas he is expounding as he does on those ideas themselves ... he presents [the exception] to show the limitations of even the worthiest rule ... and to underline again the tyranny of circumstance and the importance of keeping one's mind clear about the objective.[80]

Yet Bernard Brodie elsewhere called Clausewitz's book 'not simply the greatest but the only truly great book on war'.[81] Along with Herman Kahn, the Wohlstetters, and others, Brodie led the American 'Neo-Clausewitzian' school of the early Cold War period (see Chapter 7).

It took the Vietnam War for Clausewitz to gain the full attention of the US military. The Naval War College made *On War* compulsory reading in 1976, followed by the Air War College in 1978 and the Army War College in 1981.[82] The part of Clausewitz's work emphasising decisive battles became particularly popular and influential in US strategy in the 1980s, and in the US conduct of the Gulf War of 1990–91.

Inspired by the American Neo-Clausewitzians, interest in Clausewitz was revived in France during the Cold War: since the 1960s, this interest has manifested itself particularly on the politico-philosophical level, exemplified by the interest shown in him by philosopher André Glucksmann, the sociologist Raymond Aron,[83] the economist Christian Schmitt,[84] and other *intellectuels*, and also in journals such as the *Revue de Métaphysique et de Morale*[85] and in series such as *Philosophies*.[86] The contribution made by these French intellectuals to our understanding of Clausewitz is a little ambivalent, however. The Israeli scholar Azar Gat even argues that Raymond Aron 'is inclined to read into Clausewitz's work intellectual patterns and categories which are totally artificial and which obscure even further a subject which is already obscure enough.'[87]

In Russia, Clausewitz first became famous through his minor role in Tolstoy's *War and Peace*, where he appears in Part 10, Chapter 25, as a literary figure: here he advises a Russian colleague that 'The war ought to be carried on over a wide extent of country' as 'the object is to wear out the enemy', for which civilian sacrifices had to be accepted.[88] Clausewitz's first Russian military admirer, however, General Mikhail Ivanovich Drago-mirov, translated his works into French rather than Russian.[89] In 1897, the Polish-Russian economist Ivan Stanislavivich Bloch, who published his famous thesis that a further war among the European powers was impossible due to the huge vested interest in reciprocal commerce and prosperity, borrowed some ideas from Clausewitz. (The work was subsequently translated from the Russian into German in 1899 and into

English in 1902.) Then in 1905, General Voyde published the first Russian translation of *On War*.[90] Otherwise Clausewitz remained largely undiscovered in Russia until Lenin brought his knowledge of Clausewitz back from exile in Switzerland.

Lenin exhorted party functionaries to read Clausewitz,[91] and wrote in his 'Bankruptcy of the Second International' (Volume 21 of Lenin's complete works):

> The fundamental thesis of the dialectic . . . is that war is a simple continuation of policy by other means (more precisely, by violence). Such is the formula of Clausewitz, one of the greatest writers on the history of war, and whose thought was inspired by Hegel. This was also the viewpoint of Marx and Engels who regarded any war as the continuation of the policies of enemy powers – and of different classes within these countries in a defined period.[92]

The second and third Russian translations of *On War* were published in 1932 and 1941, the introduction to the former praising Clausewitz for his 'dialectic' approach. And the great Soviet strategists Marshals Tukhachevsky and Zhukov were both known to be admirers of Clausewitz's writing.[93] After 1945, however, as we shall see in Chapter 7, Stalin decreed that Clausewitz was not a good subject for study, and there was a temporary decline in the Soviet military enthusiasm for him.

On War was first translated into Chinese (from a Japanese translation of the German original) in 1910, and three further editions were published until 1937, when the version which is most likely to have been the one which Mao Zedong studied was published. We owe it to the work of Zhang Yuan-Lin that we now know for certain that Mao Zedong read Clausewitz's *On War*, and have a fair idea of the degree to which Clausewitz influenced Mao's thinking. In 1938, while the Japanese occupied large parts of China, Mao read *On War* in a Chinese translation, and even held a political seminar with some of his partisans on its contents, concentrating mainly on the first books of the work.[94] Between 1938 and 1976, six further versions of *On War* were published in China and Taiwan.[95] Towards the very end of his life, Mao again praised Clausewitz as a genius to the first West German chancellor who visited China;[96] as we shall see in Chapter 6, Clausewitz became fundamental to Mao's teaching on war.

In the decades before the outbreak of the First World War, Western society was generally inclined to overvalue the military and to romanticise war. In this context, military authors flourished, which accounts in part for the wide international interest in Clausewitz. Not surprisingly, however, Clausewitz was most widely read and quoted in Prussia, and, after 1871, throughout the second German Reich. As we shall see in some detail later, the military writers and generals Friedrich von Bernhardi (1849–1930), Colmar von der Goltz (1843–1916) and Hugo von Freytag-Loringhoven (1855–1924) saw Clausewitz as a source of inspiration and contributed to his popularisation. *On War* was standard reading in the military academies and widely quoted.

In the inter-war period and under the Third Reich, the National Socialists found great inspiration in Clausewitz's work, which they used selectively both to support their own views and to be seen to stand in the tradition of the Prussian martial spirit. Walther Malmsten Schering, Professor of *Wehrphilosophie* (defence philosophy) at the University of Berlin, praised Clausewitz thus:

> Clausewitz is the classic of teaching on war. In spirit he should be counted among our classical thinkers and poets . . . he stands on a level with Goethe and Schiller, with Kant, Fichte and Hegel . . . As he did not study classical antiquity and always followed his own way, the Germanic feeling of life and fate find their strong expression in him. Not as a soldier, but as a Germanic human being, he has a direct track to the concept of defence.[97]

Clausewitz was admitted into the Olympus of Great Aryans by Hitler and his followers. Hitler himself referred to Clausewitz frequently, quoted popular snippets from Clausewitz's works in his speeches and declarations, and was clearly familiar with several of his ideas.[98] Following the Führer, National Socialist authors called Clausewitz a great 'companion' of their movement, along with Johann Gottlieb Fichte, Ernst Moritz Arndt, Scharnhorst and Friedrich Ludwig '*Turnvater*' Jahn.[99] Abroad his work was called 'the blueprint from which Nazi Germany has developed the present total war'.[100] Clausewitz's anti-French patriotic 'Confession', which Hitler explicitly referred to in a speech, was read on the radio on New Year's Eve in 1944 just before the Führer's address to the German nation.[101]

When the Third Reich had turned into ashes and ruins, one German general commented that to hand *On War* to the military resembled 'allowing a child to play with a razor blade'.[102] Clausewitz nevertheless continued to be revered by survivors of Hitler's evil empire such as Hans Rothfels, who called *On War* 'The first study of war that truly grapples with the fundamentals of its subject, and the first to evolve a pattern of thought adaptable to every stage of military history and practice.'[103] In West Germany, as in France, there was sporadic interest in the more philosophical side of Clausewitz,[104] even though the military, here as elsewhere, continued to feel safer with the more concrete side of his teaching.[105]

There was barely any interest in Clausewitz in the Axis-ally Italy. The first translation of the full text of *On War* into Italian only appeared in 1942.[106] The third party to the Axis treaty, however, was long familiar with his work: Japanese staff officers first became acquainted with Clausewitz through the teaching of German officers at their military staff college from 1885–95. During that time, sections of *On War* were translated into Japanese via French, even though it was only in 1903, on the eve of the Russo-Japanese war, that the whole was translated and published. In 1904, Clausewitz's first publishing house in Germany, Dümmler, sent a copy of *On War* to General Count Tamemoto Kuroki (1844–1923), the victor over Russia in the 1904 battle of the Yalu, and already familiar with Clausewitz's main work.[107]

Further Japanese editions followed in 1907, 1913 and 1934. The latter came in the middle of a 'Clausewitz boom', which led to the publication of a series of studies of Clausewitz's main work in Japan. Indeed, the Japanese historian Yugo Asano has attributed the Japanese successes in their Manchurian campaign to the military leadership's quest for decisive battles and victory through annihilation of the enemy's armed forces, noting a specific debt to those parts of Clausewitz's teaching that had been so popular in Europe on the eve of World War I.

A new burst of interest in Clausewitz came in the mid-1960s, possibly under the influence of American strategic writing. In addition, Japanese writers took a very particular interest in the application of Clausewitz's teaching to business management. On the whole, however, Clausewitz's influence seems to have remained limited because of the notorious difficulties of rendering him comprehensible in Japanese.[108]

Since the beginning of the twentieth century, and particularly since the end of the Second World War, knowledge about and interest in Clausewitz has spread throughout most of the Western world, from Israel to Australia. The contribution of a growing number of civilians to defence issues has helped in this respect, as he enjoys more popularity with policy analysts and defence academics than with pure military practitioners, most of whom are concerned with the nuts and bolts of the military business. As a consequence, Clausewitz is quoted more often than read, let alone understood.

In 1917 the future commander-in-chief of the Reichswehr, General von Seeckt, wrote to his wife that he had heard Clausewitz quoted so often 'that one feels sick when even hearing his name'.[109] Yet this is no proof that all those who quoted Clausewitz actually studied, let alone understood his writings. The long list of Clausewitz translations and of strategists in many countries who were influenced by him must not cloud the often very shallow reading of *On War*. One might be led to believe that *On War* was the most important manual for strategists in most advanced countries, but the reality was not quite as simple.

Even though not all of the 1,500 copies of the first edition of *On War* had been sold, Clausewitz's cousin by marriage, Count Frederick William von Brühl, produced a second edition in 1853 in which he corrected spelling mistakes and made some small, but in at least one case significant, changes in the text (one of which will be discussed in Chapter 3). Again, there was a wave of acclamations, but even then the question arose, How many of those who bought the work actually studied it in depth? Wilhelm von Rüstow, a Prussian officer who later went into Swiss military service, thought that *On War* had become, along with Thucydides, a 'good for eternity'. Yet Rüstow also wrote, 'Clausewitz is cited often, but read very rarely; we ourselves have found many of his unconditional admirers who hardly noted that his main work was incomplete . . .'[110] A military expert commented two decades after the Franco-Prussian war that there was something curious about Clausewitz's influence: 'It is of almost mythical nature . . . the writings of this man . . . have actually been read much less than one might think, and yet his views have spread throughout the army and have become immeasurably fruitful.'[111] Indeed, among Germany's key statesmen, neither Bismarck nor Bethmann-Hollweg had read Clausewitz.[112]

Even though the claim was made in 1915 that 'all German leaders in the Great War' were 'disciples of Clausewitz', after the Second World War Field Marshal Ewald von Kleist famously told Captain Basil Liddell Hart: 'Clausewitz's teaching had fallen into neglect in this generation – even at the time when I was at the War Academy and on the General Staff. His phrases were quoted, but his books were not closely studied. He was regarded as a military philosopher rather than as a practical teacher.'[113]

Ulrich Marwedel rightly observed that *On War* was soon degraded to a collection of quotations which could be drawn upon to support just about any argument.[114] Taken out of context, and without the understanding of the evolution of Clausewitz's thinking which the following chapters will attempt to explain, passages from *On War* could be interpreted very much against the views held by Clausewitz once he had reached the fullness of his own understanding of war.

It was not uncommon in other Western countries to have only a secondhand knowledge of Clausewitz. Jay Luvaas pointed out that the eminent British strategist J. F. C. Fuller quoted Clausewitz, but cited only

French secondary literature, which led Luvaas to speculate that perhaps Fuller never read *On War* itself.[115] Liddell Hart thought few military men had ever read *On War*, the 'Holy Scriptures' of the military, in detail.[116] Luvaas suspected rather unkindly that Liddell Hart should be included among them.[117] Liddell Hart himself wrote, 'Misinterpretation has been the common fate of most prophets and thinkers in every sphere . . . It must be admitted, however, that Clausewitz invited misinterpretation more than most.'[118] It is quite clear that neither all of his greatest admirers nor all of his enemies had a good understanding of what makes *On War* so paradoxical in many places, and so difficult to boil down to a few simple tenets. Thus, for example, Colin Gray, an ardent admirer of Clausewitz's, writes on one page, 'Clausewitz's *On War* . . . is unique in the combination of sheer number, as well as quality and, with reservations, overall coherence, of insights'; yet four pages further on writes '*On War* is close to unreadable from beginning to end, given its lack of coherent intellectual or narrative trajectory, but it rewards browsing readers with gemlike insights at almost any random opening of its pages.'[119]

What we shall show in the following chapter, however, is that one should *not* simply browse through *On War* in a random way, but follow the evolution of Clausewitz's thought by reading the separate 'books' in *On War* in a very particular sequel.

2

Clausewitz the Idealist *vs* Clausewitz the Realist

To understand the time in which Clausewitz was writing, we have to picture the events he had witnessed, and which ushered in a new age not only in politics but also in warfare. Clausewitz tells how these changes turned the world of his childhood, the world of the Enlightenment and the *ancien régime*, upside down in a very short time. In the eighteenth-century world, wrote Clausewitz,

> Armies were paid for from the treasury, which princes treated almost like their privy purse or at least as the property of the government, not of the people. Apart from a few commercial issues, relations with other states concerned the treasury or the government, not the people . . . War thus became the business exclusively of the governments to the extent that these were isolated from their peoples and behaved as if they were themselves the state. Governments conducted war by means of the dollars in their coffers and of idle vagabonds recruited in their own or neighbouring provinces. Consequently the means available to all of them had certain limits, which their adversaries could in turn calculate both in terms of size and available time. War was thus deprived of its most dangerous side – its tendency towards the extreme . . .

He went on:

> As one knew the limits of the hostile forces one could feel safe from total ruin oneself, and as one felt the limits of one's own forces, one was forced to adopt limited aims . . .
>
> War thus essentially became a real game . . . its meaning was only a stronger form of diplomacy, a more vigorous way of negotiating, in which battles and sieges were the main démarches. Even the most

24

ambitious aimed only to acquire a limited advantage that they could use in the peace negotiations . . .

Looting and devastation of enemy territory, which had played such an important part in the warfare of the . . . ancients and even in the middle ages, were no longer regarded as acceptable to the spirit of the Age [of Enlightenment] . . . War was thus limited more and more to the armed forces themselves, not only in terms of its means, but also aims . . . only when a battle became unavoidable was it sought and given . . .

This was the situation when the French Revolution occurred. [In response to it,] Austria and Prussia resorted to their diplomatic way of war which soon proved inadequate . . . [For] war had suddenly once again become the business of the people, in this case a people of thirty million, all of whom regarded themselves as citizens . . . With its participation in the war, the whole people threw its weight into the scales, not merely the cabinet and the army. The means that could be used, the effort that could be made, no longer knew limits; the energy with which the war itself could be waged was no longer checked, and consequently the danger for the adversary was extreme . . .

[I]n Bonaparte's hand all this was brought to perfection, and this force of war, based on the entire strength of the people, wrought havoc upon Europe . . .[1]

The shock-effect of the French juggernaut of war, grinding into the ground the traditional armies throughout Europe, was the main impression under which Clausewitz wrote his first draft of *On War*.

His interpretation of war as deeply influenced by the degree of involvement of the people matched that of his tutor, Scharnhorst, who wrote, 'The source of the misfortune which affected the allied forces in the French Revolutionary War, must be sought deep within their own internal affairs (physical and moral) and those of the French nation.' And he saw the main reason for the French success was 'that the French wage war with the help of the resources of the entire nation . . . everything has been sacrificed for the continuation of war.'[2] There are similarities in Clausewitz's description of the warfare of the *ancien régime* with the analysis of the French strategist François Apolline Count de Guibert (1743–90).[3] Thus Clausewitz was not the only one to see a nexus between war and society, but it is fundamental to our understanding of *On War* that Clausewitz introduced it only during the last years of his life.

Clausewitz the Idealist

Clausewitz had, as we have seen, been a direct witness of the new era of warfare that dawned in Europe under Napoleon. Indeed, after the limited wars of the eighteenth century, Napoleon, this 'God of War', seemed to strip warfare of the constraints which had been imposed upon it by princes fighting as their own cautious *condottieri*.[4] The aim of the French Revolution and of Napoleon, by contrast, was the overthrow of the enemy's state, an uprooting of his political structure. As the Prussian philosopher Johann Gottlieb Fichte (1762–1814), one of the fathers of pan-German nationalism, put it in 1797:

> [T]he natural purpose of war is always the annihilation of the state with which one is at war, i.e., the subjugation of its citizens. From time to time a peace treaty (or really only an armistice) may possibly be concluded, because one state or both are temporarily exhausted, but the mutual suspicion remains, and the purpose of subjugation remains the same for both.[5]

Indeed, the term 'absolute war' is first found in the writings of Fichte on Machiavelli, about whom Clausewitz had corresponded with him: Fichte used it in the sense of a war of the people against their prince,[6] as witnessed in the French Revolution. Clausewitz used the term slightly differently, as we shall see, but with a similar revolutionary connotation.

Clausewitz was spurred to write *On War* by the bitter defeat of Prussia at the hands of Napoleon in 1806. Prussia's king had, since 1795, tried to avoid involvement in the French Revolutionary and Napoleonic Wars by staying neutral – hence Clausewitz's disparaging words about neutrality and his praise of Machiavelli, who discarded neutrality as a useful option. Writing in 1823–4 about 'Prussia in its great catastrophe', Clausewitz identified a number of reasons for Prussia's defeat: firstly, King Frederick William II had taken little interest in the army, as had his successor Frederick William III; the college of war, which was something like a board of governors running the army, was made up of slow old men. Among them was a minister of war, but the only department he commanded was that of supplies. The military commander-in-chief under the King (*Generaladjutant*) was chosen from an officer class that put more emphasis on elegant behaviour and knowledge of French than on talent, originality and lateral thinking. Clausewitz thought that a talented man, in the office of a proper minister of war with full powers, would have been a remedy for this problem. Secondly, the professional soldiers tended to be old – they had to serve 25–30 years before they retired. Their arms and equipment were old-fashioned, just as their training followed obsolete rules. Thirdly, the state itself ailed: the 'machinery of government' blindly pursued neutrality and peace, ignoring the danger Napoleon presented, and in turn the government did not inspire the population with a martial disposition. The 'spirit of the people' thus tended 'to avoid facing up to the danger, while incessantly invoking peace and neutrality', and the people themselves were 'unmartial and fainthearted'. Clausewitz thought that the King should have had a cabinet of good advisers who could have changed the situation entirely; instead, the men surrounding him were weak.[7]

Prussia was faced with a French state and military system at the other end of the scale, in which the whole of society and its resources were, as far as possible, mobilised for war. The wars that unfolded under Napoleon's leadership seemed to Clausewitz in their scale and horror to be almost perfect examples of *ideal* wars, of what war could be in its purest form of unmitigated violence and destruction; in his term, 'absolute' war. If one thinks of the Platonic concept of ideals and their imperfect manifestation in the real world, 'absolute' war would equal Plato's 'ideal' war, and in the real world, the Napoleonic Wars came closest to it. Clausewitz was so impressed by them (wars which almost swept Prussia off the map of Europe and resulted in the conquest of virtually the entire continent), that in his reflections, he initially wrote only about this ideal or 'absolute' form, a war completely without restraint, and with unlimited war aims. As Napoleon's *political* aims were limitless, Napoleon's *military* aims had no limitations either; accordingly, the violence seemed unlimited. Surprising though it may seem, in Clausewitz's early writings – indeed in his entire first draft of *On War*, and thus in his thinking on war during the longest part of his life – politics as a limiting, modifying, conditioning factor, played *no* part.

Instead, this is how he summarised and characterised the essence of war as he saw it in Napoleonic warfare: its purpose was 'the destruction of the inimical armed forces', as he taught the Crown Prince in 1810–12.[8] Or, as he wrote in Book I of *On War*: 'If war is an act of force to force the enemy to do our will, it concomitantly always aims at the shattering [*Niederwerfen*] of the enemy, that is to make him defenceless.' In order to make him defenceless, one had to 'destroy the armed forces, i.e. to make them unable to continue the war'; one had to 'conquer the land, because from that territory new armies could be recruited'. And if both these things had been done, the war could still not be regarded as over 'as long as the will of the enemy has not been broken, i.e., his government and allies have been persuaded to sign the peace and the people to submit'. As long as that is not the case, internal resistance or actions of the allies could rekindle the flame of war.[9] 'The annihilation of the inimical armed forces is thus always the means to reach the purpose of the engagement.'[10] Furthermore,

> While there are many ways to reach the aim of the war, i.e., the attainment of the political purpose, the battle is the only means, and thus everything is ruled by the highest law, that of the decision by arms.

Among all the possible purposes of a war, the annihilation of the inimical armed forces always dominates all the others.[11]

The engagement is a fight in which the annihilation or overcoming of the adversary is the purpose . . . How does one overcome the adversary? Always by annihilating his armed forces, through death or wounding or only to such an extent that he no longer wishes to continue the struggle. We can . . . consider the entire or partial annihilation of the adversary as the sole purpose of all engagements.[12]

'What do we mean by the annihilation of the inimical armed forces?' he asks. 'Their diminution, which must be greater than our own losses.' This diminution, this decimation of the enemy's strength, is to be achieved not solely through losses among his soldiers, but through breaking his moral forces (his morale) via a loss of order, confidence, courage, coherence and planning, and through a loss of territory and reserves.[13] Victory is defined by three elements: the greater loss of physical forces on the side of the enemy, the greater loss of 'moral forces', and his public admission that he has given up his own intentions.[14] The great decision, wrote Clausewitz, is to be brought about in the main battle (*Hauptschlacht*), a concept to which he dedicated several chapters of Book IV.[15] He postulated:

1. The annihilation of the inimical armed forces is the main principle [of war] . . . and the main way to reach the aim [of the war].

2. This annihilation of the armed forces mainly takes place in engagements.

3. Only large-scale and general engagements lead to great successes.

4. Successes are greatest if the engagements come together in one great battle.

5. Only in the main battle does the military commander-in-chief hold all the strings in his own hands . . .

[T]he main purpose of great battles has to be the annihilation of the inimical armed forces . . . the main battle is the most natural means of achieving a big, positive . . . purpose; . . . normally one is punished if one tries to avoid it for fear of the great decision . . .

The main battle is the bloodiest solution . . . blood is always its price and slaughter its character, from which the humanely [minded] military commander-in-chief flinches . . .

We do not want to hear about commanders-in-chief who win without shedding human blood. If bloody battles are a horrible spectacle, this must only lead us to respect war all the more, not, however, to allow swords to be blunted by humanitarian qualms, until there is once again a sharp sword among us which will hack off our arms at our shoulders.[16]

It was clearly the unsuccessful Prussian attempts to side with Napoleon in order to avoid destruction, the alliances of 1795 and 1812, which he had in mind, because through these alliances Prussia was ultimately unable to protect herself: war was not avoided, and Prussia's independence had to be bought with bloodshed – and that coming after the thorough humiliation, especially of its royal family, by Napoleon. Clausewitz also dismissed the division of one's armed forces and the attempt to fight the enemy in a series of separate battles as unproductive: Napoleon had taught him that the main battle was the be-all and end-all of success.[17]

The purpose of war, Clausewitz wrote in Book VI (again reflecting the reality of Napoleon's war aims), is 'the protection of one's own state and the shattering of the enemy's'. Above all, as we have seen, this meant victory over his armed forces (so that they were unable to defend themselves further) and the conquest of his territory. The former he regarded as more important; Napoleon's campaign against Russia in 1812, of which he had of course been an eye-witness, had demonstrated that one can occupy the enemy's territory, indeed even his capital, without having truly won the war, as long as the enemy's armed forces remain unconquered.[18] Once again we read in Book VII: 'The shattering of the enemy is the aim of the war, annihilation of the inimical armed forces the means, both in attack and defence.'[19] 'The annihilation of the inimical armed forces is the means to the end.'[20] The war plan had to aim at nothing less than 'the shattering of the adversary'.[21]

Thus, throughout *On War* we find the repetition of this mantra: in war, all is decided in battle, and indeed, large decisive battles are central to victory. But Clausewitz also found that the wars were won only if the enemy *literally* could not fight any further. While in the eighteenth century military commanders and their princes might decide to call off a war after a minor skirmish, Napoleon showed that he could reverse even a major defeat, by recruiting new soldiers. Clausewitz incessantly stressed the need to annihilate the enemy's armed forces because Napoleon had long proved invincible even if he had lost the occasional battle.

To sum up: basically, Clausewitz analysed only the Napoleonic wars in his earlier theoretical writing. This consisted of several manuscripts on strategy dating from 1804 and 1808, the summary of the syllabus for the Crown Prince written in 1810–12, and, crucially, Books II–VI of *On War*.[22] Here Clausewitz analysed and described only 'absolute' war, war set free from any political constraints. In all these texts dating from before 1827, political considerations are eclipsed, never appearing to have a direct influence on the conduct of war.[23] And yet, as we shall see in Chapter 5, it was the endless repetition in these texts of the need for the main battle and the annihilation of the enemy's forces that would appeal to military thinkers of the late nineteenth and early twentieth centuries.

The Turning Point in Clausewitz's Thinking

We owe our understanding of the evolution of Clausewitz's thinking to the historians Hans Delbrück and Eberhard Kessel.[24] Delbrück had understood that Clausewitz had been in the process of revising *On War* when he died. Kessel pinpointed how in 1827, after twelve years of post-Napoleonic peace, Clausewitz came to the conclusion that his analysis to date was too exclusively that of 'absolute' war to be applicable to *all* wars.

Most of *On War* dealt only with 'absolute' war, and not with the more limited manifestations of war that the world had known both before and since the Napoleonic era. 'One cannot hide the fact', he mused,

> that the great majority of wars and campaigns have come closer to a mere armed stand-off [*Beobachtungszustand*] than to a struggle for life and death, i.e., to a fight in which at least one side seeks a real decision. Only the wars of the nineteenth century have had this characteristic to such a degree that one can only use a theory that is based on this assumption. But it will hardly be the case that all future wars will have this characteristic; rather, it seems predictable that the majority will tend towards a mere armed stand-off again. This has to be taken into consideration when formulating a theory [about war in general].[25]

Why were these other wars so rarely like 'absolute' war? The answer, once Clausewitz began to think about it, was simple for somebody who had read Machiavelli's *Prince*, who had listened to Scharnhorst's lectures, or read Rühle von Lilienstern's handbook for officers of 1817–18.[26] A colleague of Clausewitz's at the military academy, Otto August Rühle von Lilienstern (1780–1847) had written,

> There is a Why? and a What For?, a purpose and a cause, at the bottom of every war and every [military] operation. These will determine the character and the direction of all activity.
>
> The individual operations have military purposes; the war as a whole always has a final political purpose, that means that war is undertaken ... in order to realise the political purpose upon which the State decided in view of the nation's internal and external conditions.[27]

Like Rühle von Lilienstern, Clausewitz realised that war was an instrument of politics, and could only unfold its full might if politics did not place restraints upon it. Thus limited political aims resulted in limited war aims, and in restraints imposed upon the way in which these aims would be pursued. As Clausewitz wrote in an (undated) manuscript, 'It is a great error [to think] that war is an independent thing, which has to be judged according to its own laws, in which the political elements have to be seen as an anomaly. Rather, war is nothing but politics.'[28]

In December 1827, Clausewitz articulated his new thinking in a letter. He wrote that the 'final aim of the entire military act . . . is the most important and paramount [aim] that the strategist has to establish', and this aim could take many forms, not only the unlimited one of Napoleon that had been the subject of Clausewitz's deliberations so far. .

It is obviously quite different if I have the intention . . . to strike down [*niederwerfen*] the adversary, to make him defenceless and to force him to accept my conditions for peace, or if I have to make do with acquiring an advantage through the conquest of a small piece of territory, a fortress etc., either in order to keep them when the peace [is concluded] or to offer them up in a bargain. The extraordinary conditions in which Bonaparte and France [operated] since the Revolutionary Wars have made the first option possible for him virtually always and everywhere. Therefore one has imagined that the resulting battle plans and their execution were the general norm.[29] But that would amount to disregarding the entire history of earlier wars. This would be folly. If we want to deduce the art of war from the history of war, and that is indisputably the only way to get there, we must not dismiss as unimportant the manifestation of war in history. Thus if we perhaps find that among 50 wars, 49 . . . have a limited aim and do not aim to strike down the enemy, we must understand that [the 49 cases are also] part of the nature of [war] and we must not assume every time that [these limited aims and the resulting conduct of war] are due to mistaken ideas, or a lack of energy, etc. We must therefore not allow ourselves to be tempted to regard war as an act of pure violence and annihilation, and to make a number of logical deductions from this simple concept, which do not actually match the phenomena of the real world. Instead, we must return to the [idea] that *war is a political act*, which does not bear its law within itself, [but] is a *true political instrument*, which does not act on its own, but *is guided by . . . politics*. The more politics serves great interests which are all–encompassing and existential, the more the question is one of mutual survival [lit.: 'the to be or not to be'], the more politics and enmity coincide, the more the former melts away into the latter, the simpler war becomes, the more directly it is a function of force and annihilation, the more it fits all the postulates that can be deduced logically from these two concepts . . . Such a war appears quite apolitical, and for that reason has been regarded as the norm . . . But apparently the political principle is not absent here any more than in other wars, only it coincides so entirely with the concepts of force and annihilation that it becomes invisible to our eyes.

[By contrast there can be] wars with much more limited aims, for example a mere threat, armed negotiations, or in the case of alliances, a merely symbolical gesture. It would be quite unphilosophical to pretend that these wars do not come under the rules of the art of war. Once the art of war is compelled to concede that wars can be rational which do not aim for the extreme, i.e., the shattering [*Niederwerfen*] and annihilation of the enemy, the art of war must condescend to embrace all sorts of gradations which might be required by the interests of politics. The duty and the right of the art of war in relation to policy-makers [*Politik*] is mainly to prevent policy-makers from asking for things which are contrary to the nature of war, lest they make mistakes in the wielding of this instrument through ignorance of its effects.[30]

Clausewitz realised that he had to rewrite all the earlier parts of *On War*. War was a function of politics, and almost everything he had said so far had left out this key variable. So Clausewitz finished the work with Books VII and VIII, which already reflect this realisation, and started rewriting the other six. Tragically – and in a very real sense, as we shall see – he never lived to complete this rewriting. Clausewitz was satisfied only with the rewritten Book I, as he wrote to a friend towards the end of 1829, around the time when he last worked on the book.[31]

In sum, then, we are left with two different sets of Clausewitzian teaching: on the one hand we have the draft papers of 1804 and 1808, the instructions to the Crown Prince, other historical pieces, and Books II–VI of *On War*, all dating from the years before 1827, all the works of 'Clausewitz the Idealist'. On the other hand there are Books VII, VIII and the revised Book I of *On War*, written between 1827 and Clausewitz's final posting in 1830, which are the writings of 'Clausewitz the Realist'. Unfortunately, confusion between the two mindframes persisted through-out these later books, even though he tried to revise his previous thinking. So even here we find numerous examples of his idealist thinking, mainly centring on the need to annihilate the enemy's armed forces.[32] Here is a passage which in its inconsistency reflects the tensions between Clause-witz's two different views of war:

The shattering of the enemy is the aim of war, annihilation of the inimical armed forces the means, both in attack and defence . . . Aggression . . . leads to the conquest of the country; this is the objective – but it need not be the whole country; it may be limited to a part, a province, a strip of territory, a fortress, etc. Any one of these can be of sufficient value as political cards in peace [negotiations], either to be retained or exchanged.

The object of strategic attack, therefore, may be thought of in countless gradations, from the conquest of a whole country to that of an insignificant patch of territory.[33]

How could Clausewitz state dogmatically that 'The shattering of the enemy is the aim of the war', requiring the 'annihilation of the inimical armed forces', if two lines later he realised that the objective could be anything on a wide scale of political objectives? If only an insignificant patch of territory is to be conquered in time for political negotiations, surely it is not necessary to destroy the enemy's entire fighting forces to do so. This passage thus shows the continued coexistence of his two different mind-frames.

Clausewitz the Realist

The determination of war aims by politics became the major theme in the writing of Clausewitz the Realist, dating from 1827 to 1830. In the following passage Clausewitz used the imagery of two opposing armies as electric poles:

Most wars seem little more than mutual indignation, with everybody taking up arms in order to defend themselves and to frighten their opponent and – occasionally – to deal him a blow. Thus we do not find the clash of two mutually destructive elements, but electrical tensions between separated elements, which discharge themselves in individual small shocks.

But what then is the non-conducting dividing wall, which prevents the total discharging [of energy]? Why is no justice done to the philosophical [i.e., the ideal] idea?[34]

But other wars had had that 'non-conducting' wall between the two poles, preventing an 'absolute', perfect war, and that wall consisted of the political restraints imposed upon the war.

Thus Clausewitz wrote in a letter of 22 December 1827:

War is not an independent thing but the continuation of politics with altered means. Consequently the main outlines of every major strategic plan are *primarily political in nature*, and increasingly so, the more they embrace war and the state in their entirety. The whole war plan results directly from the political characters of the two warring states, as well as from their relations to third powers. A plan of campaign is deduced from the war plan, and frequently – if everything is concentrated on one theatre of operations – may even be identical with it. But even the individual parts of a campaign contain the political element; hardly any important act of war such as a battle, etc., will be free from its influence. According to this point of view, there can be no question of a *purely military* evaluation of a comprehensive grand strategy, or of a purely military plan for it . . .

Until now, one has sought to divide the military element of a great strategic plan from the political, trying to regard the latter as something irrelevant. War is nothing but a continuation of political endeavour with altered means. I base the whole of strategy on this tenet, and believe that he who refuses to recognise this necessity does not fully understand what matters. This principle explains the whole history of war, and without it, everything would appear quite absurd.[35]

What then, if politics predominate, are the war aims that are dictated by policy? We have already seen that they can range from the occupation of an insignificant place to the conquest of a country. In Book VIII, Clausewitz expands on the way a war is planned if the aim is limited, and on the way it is planned when the aims are unlimited. He almost falls into a new trap, that of thinking of war dualistically, as one of merely two possible forms: limited or unlimited. Even back in 1804, in a piece on strategy, he had briefly touched on the 'dual form' of war: 'The political purpose of war can be of two sorts: either to annihilate the adversary entirely, to eradicate his existence as a state, or to dictate the conditions for the peace [treaty].' But then he had still thought that 'In both cases it must be [our] intention to paralyse the inimical forces to the extent that he [the enemy] cannot continue the war or can continue it only at the risk of his whole existence.'[36] In other words, back in 1804 Clausewitz had not yet believed that the conduct of a war might vary as a function of its political aims. In a note of 10 July 1827, however, in which he announced that he would have to revise Books I–VI of *On War*, he again invoked the 'dual nature of war'.[37] This curiously dualistic notion, which some took to be influenced by Hegelian dialectics, almost led Clausewitz into the logical dead end of thinking of war *only* as one of two possible manifestations. His own logic, however, led him to acknowledge elsewhere[38] that wars have to be classified along a sliding scale, from defensive but unlimited – e.g., the Russian position in 1812 – via very limited – e.g., sabre rattling on both sides as a mere gesture before getting back to diplomatic negotiations – to an all-out, unlimited, Napoleonic offensive war of conquest. In a picturesque statement of this, Clausewitz called war a 'true chameleon, because it changes its nature a little in each concrete manifestation'.[39]

In writing Books VII and VIII and revising Book I, Clausewitz used some crude and unconvincing stylistic ploys in order to make this sudden break in his thinking fit his existing text: 'We might have posited' that war is a function of politics, and that therefore there can be very different forms of war '*to begin with* [my italics], if it had not been necessary to emphasise the contradictions with all possible clarity and to consider the different elements separately.'[40] But he did realise that these stylistic ploys were insufficient, that he really had to start all over again.

Books VIII and I are thus full of passages in which he reiterated the same principle: 'War is only a part of political activity, i.e., in no way autonomous.'

> While it is widely understood that war is caused by the political intercourse between governments and people, it is generally thought that with the beginning of war, that intercourse ceases and quite a different situation arises which is subject only to their own laws. We, however, contend that war is nothing but a continuation of political intercourse with the addition of other means ... This political intercourse does not come to an end because of war, nor is it changed into something quite different, but it continues to exist in its essence, whatever form the means might take which it uses ... Thus war can never be separated from political intercourse ... War ... cannot follow its own laws, but has to be seen as part of a whole, and the whole is politics ... [P]olitics ... turns war into a mere instrument ...
>
> As war is part of politics, it will take on its character. As soon as politics is great and mighty, so it will be for war, and this can be the case to the degree where war assumes its absolute form.[41]

> Once again: war is an instrument of policy;[42] it must necessarily bear the character of policy and has to apply its standards; the leader of war, in its great outlines, is therefore policy itself, laying down the pen and taking up the sword instead, but policy does not on that account cease to think according to its own laws.[43]
>
> We thus see first, that under all circumstances, we have to think of war not as something independent, but as a political instrument; ... Secondly, this conspect shows us how different wars have to be, depending on the nature of the motives and situation from which they have arisen.[44]

Indeed, the aims of any war, the political situation which it is waged to achieve, influences the very way it is conducted.[45]

The smaller the sacrifice we demand of our adversary, the smaller we can expect his efforts to be to deny it to us. The smaller his efforts, the smaller ours can be. Moreover, the smaller our political purpose is, the lesser the value is that we accord to it, and the sooner we will give it up, or, for the same reason, the smaller our efforts will be to fulfil it. Thus the political purpose as the original motive for the war will be the measure both for the aaim that has to be reached through the action of going to war, and for the efforts that are necessary.[46]

Or, as he formulated it a few pages on,

The greater and the stronger the motives of a war, the more they affect the entire existence of peoples. The more violent the tension that leads to the war, the more war will approach its abstract [ideal] form, the more it will aim to shatter the enemy. The closer the confluence of military war aim and the political purpose, and the more purely military, the less political the war appears to be. The weaker the motives and tensions ... the more war will be deflected [by the political leadership] from its natural tendency [towards extreme violence]. The greater the difference between the political purpose and the aims of an ideal war, the more *political* the war appears to be.[47]

But, he added insightfully, the same political purpose can have a very different value for two different peoples, or indeed even for the same people at two different times in its history: coming on top of a wide array of other considerations, it might either be stifled by the preponderance of factors moving a people to keep the peace, or it might serve to ignite a powder keg of other factors driving the people to go to war.[48]

War waged by a community, by entire peoples, and especially by civilised peoples, always takes a political situation as its point of departure and is called into being exclusively by a political motive. It is thus a political act.

For that reason, instead of being an unrestrained explosion in all directions, war remains subject to restraining factors, to friction, but also and more importantly to political control: it will remain subject to the intelligence which is directing it.

If we consider that war emanates from a political purpose, it is only natural that this prime mover which has called it into being will remain the first and highest concern in its conduct. Yet the political purpose is not a despotic legislator: it has to conform to the nature of the means and in the process of adapting to it may become significantly transformed. It will nonetheless remain the first consideration. Politics will thus permeate the entire act of war and will have a continuous effect upon it, to the extent that the nature of the exploding forces of war permit.

War is a mere continuation of politics with other means.

Thus we see that war is not merely a political act, but a true political instrument, a continuation of political intercourse, an execution of policy with other means. All that remains particular to war is the particular nature of its means. The art of war in general and the military commander in every particular case can insist that the directions and intentions of politics do not conflict with these [military] means . . .

The nature of war, the specific military tools available to a government, will force it to adapt its aims and will thus have a certain influence on the political purpose that the government has adopted. Nevertheless, the policy remains the dominant factor, the resulting military acts are a function of this factor, and to understand any war one has to understand both its political purpose and the nature of the military tools through which it is conducted.[49]

Real war is thus almost always less violent than ideal war. If abstract, ideal, absolute war aims to render the adversary defenceless, this is not necessarily the aim in all war, as Clausewitz the Idealist had claimed so often. In real wars, in real history, there were countless examples of a peace concluded without 'shattering the enemy' or bringing down his state. Not every war needs to be fought to the finish, to the collapse of one side or the other. 'One can imagine that in case of very weak motives and tensions' a small hint that one might resort to the use of force, a veiled threat, might persuade the other side to yield.

Perhaps over-optimistically, Clausewitz postulated that 'war is no act of blind passion' (not a tenet he could have maintained after the experience of the two world wars of the twentieth century). From this tenet he deduced that the political purpose of any war and its value to the government must determine the size, length and intensity of the effort one would make to obtain this purpose. If the effort weighs too heavily compared with the political purpose it is intended to serve, the latter has to be given up, and one has to conclude peace.[50] (Again, this reasonable proposition was not to find universal application in the following century, if one thinks of the price paid for the achievements of the First World War.)

In his rewriting of Book I, Clausewitz went on to discuss a series of ways in which force could be used. No longer did he see the annihilation of the enemy's armed forces and the occupation of his territory as the *only* options. One might opt for a series of battles and victories until the enemy was crushed, but one might content oneself with one spectacular victory which would shatter the enemy's self-perception of invincibility and inspire him with a lasting respect for the victor, which might rule their relationship instead of further contests between the two sides. Or else one might weaken the adversary decisively through bloodless means, for example by seducing his allies into deserting him, or by enlisting further powers on one's own side which would change the balance of forces so decisively that no battle was needed to prove one's superiority, and the adversary might give in without bloodshed.[51] As the effort that both sides have to make in war can be measured in the dimensions of size, intensity and time, and are a function of the overall motive or political purpose, Clausewitz saw that an imbalance in one of these dimensions might be offset by the inverse imbalance in another dimension. Clausewitz concluded that

> in war there are many ways to one's goal which is not in every case the shattering of the enemy. The annihilation of the inimical army, the conquest of inimical provinces, their mere occupation, their mere invasion, actions which aim to change the political relations between the adversaries, finally passive waiting for the enemy to strike, these are all means which, each for itself, can be used to overcome the enemy's will, depending on the peculiarities of every case . . .[52]

And a little further on he has plainly inserted new text into Book I. We still find Clausewitz the Idealist stating, 'The annihilation of the inimical armed force is thus always the means to reach the purpose of the engagement', followed by Clausewitz the Realist adding:

> This purpose can be the mere annihilation of the inimical armed force, but this is not always necessary and can be something quite different. For as soon as . . . the shattering of the adversary is not the only means of reaching one's political purpose, as soon as there are other war aims that one can adopt

in pursuit of that purpose, then these can substitute for the annihilation of enemy forces.

As an example he cited the military aim of taking a mountain or a bridge. In such a case the annihilation of the enemy forces would be as valid a way to proceed as it would be merely to chase away the enemy's soldiers who have been holding that strategic point. He even conceded that a battle might become unnecessary if the imbalance of the forces between the two sides was so great that the weaker would be persuaded to give in on the basis of his mere estimate of the outcome. Yet – and here it is that he found the synthesis of his two contradictory tenets – the basis of all conflicts is the battle; the *actual* battle or the *virtual* battle in a comparison of forces. Even the battle that is called off at the last minute, with the weaker side making important concessions without having lost the field, has to be credibly threatened to have this effect. And this is where he repeated his famous tenet dating back to 1804 that 'The decision by force of arms is for all large and small operations of war what cash payment is in commerce'.[53] Any diplomatic threat must be backed by the credibility and likelihood of its military implementation just as a cheque has to be backed by credit in the bank. If the use of force is not threatened credibly, the bluff might be called and exposed as such.

In Book VIII, Clausewitz lets us share his intellectual development, giving the historical overview of the development of warfare which, really, he should have given at the very beginning of *On War*. I have quoted much of it at the beginning of this chapter, and he sums up thus:

> Half-civilised Tartars, republics of classical Antiquity, feudal lords and trading towns of the middle ages, kings of the eighteenth century, finally the princes and peoples of the nineteenth century – all wage war in their own way, wage it differently, with different means and with different purposes.[54]

He explains that he has given this overview

> not in order quickly to assign a few principles of warfare to each period, but in order to show how every age had its own kind of war, its own limiting conditions, and its own preconceptions. Each [period], there-fore, would have its own theory of war, even if one had been inclined everywhere and at all times to deduce it from philosophical first principles. The events of every age must therefore be judged in the light of its own peculiarities ...
>
> But the conduct of war, though conditioned by the particular characteristics of states and their armed forces, must surely contain some more general – indeed a universal – element with which theory should be concerned.
>
> Recent times, in which war has attained its absolute force, show most of the general and necessary characteristics. Yet it is no more likely that war will always be so monumental in character than that the barriers which have been broken [in the Napoleonic Wars] will be re-established. A theory, then, that deals exclusively with absolute war would either have to ignore any case in which the nature of war had been changed by alien influences, or else it would have to dismiss them all as mistakes. That cannot be the purpose of theory. Theory must therefore ... always bear in mind the wide variety of circumstances in which war

comes into being. Theory will then draw the outlines of its salient features in such a way that it can accommodate both the dictates of the age, and of the moment.

In conclusion we must say that the aims a belligerent adopts, and the resources he employs, must be governed by the particular features of his own situation. But they will also conform to the character of the age and to its general situation. Finally, they must always be governed by the general conclusions to be drawn from the nature of war itself.[55]

Clausewitz's historical survey in Book VIII gives us so many examples of limited war, that he feels forced to write, 'One could doubt that our idea of the absolute nature [of war] has any reality, had we not precisely in our own time seen real war appear in its absolute perfection. After a short introduction by the French Revolution, the reckless Bonaparte has brought it to that point [of perfection].'[56] (And he repeated, as we have already seen, that it was 'Recent times, in which war attained its absolute power . . .'[57]) The phenomenon of these wars 'naturally and necessarily' led him to the discovery of the 'original concept' [*ursprüngliche Begriff*] of war, to the ideal of war. But in writing a theory about it, Clausewitz had to face the question 'whether war should only be' like the Napoleonic Wars, 'or whether it can be different'. And having found so much evidence of the latter in history, he dismissed the easy option which had dominated his earlier work, namely to find a clear, simple theory that fitted only the Napoleonic Wars but was ill-suited to explain what had happened in so many other wars. Instead, the theory he wanted to establish had to fit the diverse historical evidence:

War, as it should be, is to be constructed not merely as its [ideal] concept, but we have to concede space to all the alien elements which are mixed together with it, all the natural inertia and friction, the lack of logic and clarity, the despondency of the human spirit. We have to concede that war . . . is the result of ideas, feelings and situations which immediately precede it. Indeed, if we want to be meticulously right, we have to concede that this was even the case where war has assumed its absolute form, namely under Bonaparte . . .

All this has to be admitted by theory, but it is the duty of theory to put the absolute form of war above everything else and to use it as the general point of reference. In this way he who wants to learn from this theory shall . . . never lose sight of it, shall regard it as the original measure of all his hopes and fears, to come close to this theory, wherever he can and is forced to.[58]

Speculating about the future of war, he pondered:

Whether it will always stay like this, whether all future wars in Europe will always be fought with the total weight of the states and therefore only about great interests which are close to the peoples' hearts, or whether gradually there will once again be a division between government and people, is difficult to decide . . .[59]

The implication therefore is that when the peoples' 'great interests' (what would later be called 'vital interests') are at stake, you have 'absolute' wars, while when governments fight wars without harnessing popular support, they are fought for limited war aims, and are themselves limited. This is a very important point. (Indeed, it is striking that there were no further 'absolute' wars, with the mobilisation of the total population akin to the French *levée en masse*, for about half a century.)

The Two Clausewitzes

Even after his great insight of 1827, Clausewitz regarded absolute war as more true to the essence of war (he uses the term *Begriff*), and merely finds in politics, friction and circumstances the reasons why a particular war stops short of becoming absolute war.[60]

We have to remind ourselves, in conclusion, that Clausewitz not only did not complete the revision of his work, but that he was only happy with the sole revised book, Book I, by the time he stopped working on the *opus*.[61] To recapitulate, in his earlier work – the first draft of Book I and Books II–VI – Clausewitz wrote mainly with 'absolute' war in mind, war that unfailingly aims to shatter the enemy in a major, decisive battle, to bring down his state, occupy his territory and break his will to resist. In Book VIII he described it as the wars he experienced in his time, wars that Napoleon brought 'to their absolute perfection'. In the first version Book I he wrote of 'absolute' war more as of an ideal, a theoretical concept (*Begriff*), which reality can come close to but which mostly remains an abstract, the extreme idea of what war could be like. In Books VII, VIII and the revised Book I, written after 1827, Clausewitz extended his theory of war to encompass not only 'absolute' war but the many more limited manifestations of war that he had found in history. Only here did he introduce the famous tenet of war as a function of the political purpose for which it is waged. Thus Clausewitz only introduced the political variable, for which he is so famous, into his work in the last four years of his life.

Clausewitz's work, although unfinished, and even though it lent itself to misinterpretation with most devastating consequences, is outstanding in its strength and even in its confusion. For even in his initial blindness to the importance of politics in war, Clausewitz understood something very important: namely, the boundless horror of war and its potential to 'explode', to get out of hand, due to its intrinsically violent nature, even when the political purpose is clearly defined and discipline prevails in the armed forces. The tension between his 'absolute' war and the limited wars he knew is the contradiction between the unleashing of brutality, and self-discipline; between the break-down of the laws of human society (chief among them 'Thou shalt not kill'), and the imposition of restraints on the ensuing chaos and slaughter. Even in his contradictions, he shows great wisdom. As we have seen he repeatedly claims that the total annihilation of the enemy is necessary for a real achievement of the war aims and that the enemy's country must be occupied, 'for otherwise new forces can be created from that land'; on the other hand he notes that, as politics dictate warfare, a war must be regarded as well and truly finished once peace is made.[62] Realistically, looking at the experience of the Allied victory over Germany in the First World War, for example, after which Germany was not occupied, or looking at the experience of the Gulf War against Iraq in 1990–1, after which, again, Iraq was not occupied, one can see that in his contradiction, Clausewitz is still right on both counts. The tensions and contradictions in his unfinished work are largely due to the conflicts between realities and intentions in the real world.

Captain Sir Basil Liddell Hart was thus right up to a point in his criticism of Clausewitz. He saw that there was a contradiction in Clausewitz's writing but he did not resolve the riddle. He recognised the passages qualifying the tendency of war to go to the extreme with the political aims that influenced it, but added:

Unfortunately, his qualifications came on later pages, and were conveyed in philosophical language that befogged the plain soldier, essentially concrete minded. Such readers grasped the obvious implication of the leading phrases, and lost sight of what followed owing to distance and obscurity. In justice to Clausewitz one must draw attention to his reservations, but for true history one must concentrate attention on his abstract generalisations, because it was the effect of these that influenced the course of European history . . . For it was the ideal, and not the practical [what we have here called realistic], aspect of his teaching on battle which survived . . . Not one reader in a hundred was likely to follow the subtlety of his logic, or to preserve a true balance amid such philosophical jugglery. But everyone could catch such ringing phrases as – 'We have only one means in war – the battle.' 'The combat is the single activity in war' . . . 'The bloody solution of the crisis, the effort for the destruction of the enemy's forces, is the first-born son of war.' 'Only great and general battles can produce great results.' 'Let us not hear of generals who conquer without bloodshed.'[63]

Probably most students of Clausewitz shared the lack of understanding which Liddell Hart here attributed to the average reader. Hence we mostly find a very one-sided reading of Clausewitz and the claim, made explicit or implied, that there is just one Clausewitzian model of war, when in fact there are at least two main permutations, the wars of Clausewitz the Idealist and the wars of Clausewitz the Realist.

Nor is it right to regard these two views of war as mutually exclusive. It is no more so than for the physicist to think of light in terms of waves or particles. Just as scientists have forever been looking for the one formula which explains the basis of all natural phenomena, strategists have long been mesmerised by the quest for a simple set of rules summarising all strategic knowledge, along the lines of the colloquial soldier's exhortation to 'hit the enemy the hardest the fastest where it hurts the mostest, when he expects it the leastest'. But on a more abstract level, no such formula has been found other than the Clausewitzian realist formula that war is a function of the variable 'politics'. Yet every war contains the potential for escalation, more or less effectively checked by political considerations, or more or less effectively amplified, again by political considerations. Both the teachings of Clausewitz the Idealist on 'absolute' war, and of Clausewitz the Realist on war as a function of politics, apply to the phenomenon of war, but they must be applied discriminatingly, with an eye to all the variables involved.

For the views of war of Clausewitz the Idealist and that of Clausewitz the Realist are not all there is to Clausewitz's thinking. In the following chapter we will turn to his elaboration of the role of politics in warfare, and the relationship between war and society, before looking in Chapter 4 at some of the more technical discoveries for which Clausewitz is famous, and the echoes of his insights found in later authors' writing.

3

Politics, the Trinity and Civil–Military Relations

Clausewitz was neither the first to recognise the importance of politics in warfare, nor to recognise the political dimension of several other aspects of war. In this chapter we will explore the extent to which Clausewitz was following others, and the extent to which he in turn influenced others. We will begin by looking at war as a tool of politics, and then focus on Clausewitz's view of the world, of society and of relations between states. Finally, we shall illuminate Clausewitz's thinking about the relationships between what we would today call key policy-makers (whom Clausewitz subsumed under the heading of *Politik*) and the military commander-in-chief (*der Feldherr*). As we shall see, this was a much debated issue, and the misrepresentation of Clausewitz's view on the subject was of considerable consequence.

War as a Tool of Politics

Machiavelli took for granted the idea that war is a tool of politics, and Napoleon himself said in one of his maxims, 'Political reason is above everything else.'[1] And inspired by the clear nexus between the political aims of the French Revolution and the new French way of war, several of Clausewitz's contemporaries hit on similar observations. F. Constantin von Lossau, who, like Clausewitz, was a Prussian military officer and took part in the wars against Napoleon, preceded Clausewitz in his reflections on the relationship between politics and war. In his own book, *Der Krieg* ('War') of 1815, he called war 'a most extreme tool of the states':

> Politics ensures the protection and the prosperity of the state, it defends the [state's] individual interest; [politics] indicates the basic idea, the

direction and the aim which the state is to pursue. Where [politics] ceases to have an effect, war begins.

And politics, he argued, provides the purpose of the state, while war is the means for its pursuit.[2] Thus, like Clausewitz, he saw the nexus between war and politics, but unlike him he claimed that politics operated only outside war.

Otto August Rühle von Lilienstern, who was the same age as Clausewitz, and like him had followed Scharnhorst's courses and taught at the War Academy, disagreed with Clausewitz on the education of the officers who were put in their charge. Lilienstern thought they should be taught a wide array of subjects to give them greater general knowledge. Clausewitz, by contrast, thought that the officers should concentrate on learning what was of direct importance to their occupation, just as he kept his own *On War* free from any debates about politics and the political purpose of war.[3] In 1817–18, 'R. von L.' published his *Manual for Officers*, in which he wrote that while the individual operations of a war have a military purpose, the final purpose of war is always political in nature. War is the tool of statecraft which is used to achieve political purposes:

> War is . . . the means of states to assert their rights or wrongs, in other words, their political purposes against each other; and the realisation of these political purposes is the true final purpose of war, not victory, peace or conquest, unless these happen to fit the political intention.[4]

Clausewitz's thinking is clearly very close to Rühle von Lilienstern's, but while Clausewitz was thus not the only one to think of war in connection with politics, it was his work which conveyed these ideas to subsequent generations, while von Lossau and Rühle von Lilienstern were forgotten in time.

The Importance of Politics to Clausewitz's Followers

Oddly, few of Clausewitz's readers initially took an interest in the nexus between politics and war that he described. In a letter to Marx, Friedrich Engels commented on the passage in *On War* in which war is described as a social activity: 'I am currently reading Clausewitz, amongst others . . . To the question whether we should be talking about the art of war or the science of war, we find the answer that war most resembles commerce. The engagement is in war what cash payment is in commerce, however rarely it has to occur in reality, all turns on it, and in the end it has to take place and is decisive.'[5]

Lenin studied *On War* while he was in exile in Switzerland. He noted down excerpts with his own commentary, showing particular interest in Books I and VIII. He was impressed by Clausewitz's treatment of war as a social phenomenon, as something that could mobilise the interests of the entire society and would take a wrong turn if it served primarily 'the ambitions, the private interests, the vanity of the members of the government'. He was intrigued by Clausewitz's thoughts on the relationship between politics and war, noting for himself that 'wars . . . are not conducted alone because of enmity'. He espoused the notion that war is a function of politics, but he had difficulties with Clausewitz's idea that the more political a war was, the less violent it was, noting: 'semblances are not reality. War seems the more "bellicose", the more deeply it is political, and the more "political", the less deeply political it is.' He developed Clausewitz's ideas further, noting that the same politics that dominated war continued into peace time, and that the particular character of a society (e.g., Revolutionary France) not only influenced its conduct of war but determined also whether it posed a threat to its neighbours. Accordingly, a conflict is not always over for ever when a peace is concluded, as the same political and social forces which brought about the war in the first place might still be there. Lenin underlined Clausewitz's conviction that the defeated party had 'the right to a resurrection'.

Lenin was deeply impressed by Clausewitz's recognition that each period had its own form of war.[6] In April 1917 he declared, *ex cathedra*, 'that Marxists see in war the continuation of politics which are conducted by certain governments as representatives of certain classes.' It was in this context – of a comprehensive critique of Plechanov – that Lenin wrote, 'War is the continuation of politics . . . War and politics are connected with the interests of certain classes, and . . . one has to examine which classes are waging the war and for what reason they are waging it.' Soon thereafter, in a speech in Petrograd, he articulated his famous maxim:

> *War is the continuation of the politics of one class* [my italics], and to change the character of war means to change the class that holds power . . . Our party will patiently . . . try to explain to the people the truth that . . . a war . . . can only be terminated by a democratic rather than by a violently [dictated] peace if the entire power of the state is transferred into the hands of that class . . . which is really capable of terminating the oppression at the hands of the capital.[7]

And in his speech of May 1917 in Moscow ('War and Revolution'), Lenin explained:

There are wars and wars. One has to examine the historical conditions from which every individual war has arisen, which classes conduct it and with what aim. If we do not do so, all our deliberations about war are nothing but hot air . . . With regard to the subject of the relationship between war and revolution . . . the dictum of Clausewitz, one of the most famous writers on the philosophy of war and the history of war, is well known, which is: 'War is a mere continuation of politics with other means' . . . This writer, whose basic ideas are today shared unconditionally by every thinking person, even eighty years ago fought against the prejudice of philistines and ignorants that war could be detached from the politics of the respective governments, the respective classes, or that one could simply regard war as a mere attack that is a breach of the peace, which is then simply restored. To row and make up! That is a primitive notion, based on ignorance, which has been proved wrong decades ago and can be contradicted by any half decent analysis of any historical era of wars.

War is a continuation of politics with other means. Every war is insolubly linked with the political order from which it has sprung . . .[8]

Imbued with the Leninist spirit, in 1927 the Soviet strategist A. A. Svetchin began to extol the contribution made by Clausewitz to strategic thinking, terming it a 'Copernican revolution' compared with earlier strategic thinkers like Bülow and Lloyd.[9]

Standing in the same tradition, Mao Zedong also keenly took up Clausewitz's ideas: 'What is said with the sentence "war is a continuation of politics" is that war is politics, and that war itself is action with a political character; since the beginning of time there has been no war which has not had a political character.' Nevertheless, war for Mao was something special, and 'cannot simply be equated with politics in general'. From which he concluded, 'Politics is war without the shedding of blood, and war is politics with the shedding of blood.' This in turn meant that there is 'a whole system of specific organisations of war, a whole series of specific methods of the conduct of war, a specific process of war'. And he developed this theme in the context of the Japanese occupation of China in the 1930s and during the Second World War:

If politics has developed to a certain stage where it can no longer be continued in the old way, then war breaks out with the help of which the obstacles in the path of politics are brushed aside. Thus the semi-independent status of China was an obstacle in the development of the politics of Japanese imperialism, and thus Japan unleashed the aggressive war in order to sweep this obstacle aside. And what of China? The imperialist yoke has long become an obstacle for the bourgeois-democratic revolution in China, and that is why many wars of liberation have occurred in China with the aim of eliminating this obstacle. As Japan is now applying the means of war in order to oppress China and to bar the way of progress to the Chinese Revolution, we are forced to wage a war of resistance against Japan and with determination to remove this obstacle from our way. As soon as obstacles are moved out of the way and the political aim has been attained, war comes to an end. But if the obstacles have not been removed entirely, the war has to be continued until the aim has been reached completely.[10]

Mao took from Clausewitz the lesson that one must 'not separate war from politics for a minute'.[11]

Similar ideas were echoed at the other end of the political spectrum, where extremes meet. The Fascist and National-Socialist-sympathising state philosopher and lawyer Carl Schmitt (1888–1985) adapted Clausewitz's thinking in a similar direction. He linked politics with war by seeing in politics the eternal potential of strife and thus of war. Commenting on Clausewitz's vision of the relationship of politics and war, Schmitt saw politics (including domestic politics) as dominated by the potential of armed conflict which at once sharpened the political disputes and disciplined them. Writing later during the Cold War, he criticised Clausewitz and indeed international law for focusing so much on inter-state war, which he called a 'conventional game' compared with true war, war inspired by intense hatred, which he thought one could find only in precisely those two forms of war which are treated as marginal in international law: civil war, and anti-colonial war. Clausewitz in the last years of his life denied the need for physical annihilation of every last soldier in the enemy's armed forces, insisting on the importance merely of changing the enemy's mind; for Carl Schmitt, by contrast, the desire to annihilate the opponent was intrinsic to his definition of true war.[12]

The French philosopher Michel Foucault echoed Lenin by describing politics as the continuation of war: 'We are thus all continually and permanently in a state of war against each other; a battle front cuts right across the entire society . . . It is thus a front which puts each of us in one camp or the other. There is no neutral subject. One is necessarily somebody's adversary.'[13] However, as another Frenchman, Emmanuel Terray, commented,

Clausewitz states that there is a continuity and a similarity between politics and war, but, by politics, he means above all diplomacy. If the internal life of States also resembles a state of war, this can only be the effect of a deplorable derailing [of society]. On the basis of this conviction, he can maintain that there is still a clear distinction between war and society. By contrast, those thinkers who see conflict as the normal mode of political existence – within the States as much as between them – are naturally inclined to deny or to reduce that distinction. One can argue that they thus move far away from the views of Clausewitz; it is nevertheless Clausewitz who has opened that door for them.[14]

Clausewitz's World-View

On War presupposes a very particular world order, namely one in which the state plays the most important role, even though in his historical survey Clausewitz recognised that in different ages, states had a differing degree of cohesion, and that even within the same age, states that were republics could not go to war and wage war as easily as some autocracies.[15] In Book VI (Chapter 6, Section 6) of *On War*, Clausewitz outlined his view of the inter-state system of Europe. He denied that there was a systematically regulated equilibrium of power and interests, but there was nevertheless a system in existence, which he pictured as a network of conflicting interests, in which every knot symbolised the focal point of conflicting demands of different states in one matter, each state pulling its string in a different direction, so that there was a multitude of balanced strains in this network. Any attempt by one state to change this distribution of forces would be resisted by all others, so that the overall system had the tendency to resist change, thereby serving the interest of the majority. Clausewitz conceded that the system had at times failed to match the force and determination of a single state that set out to dominate Europe, but maintained that the system's tendency to work for the collective interest always re-emerged in due course. 'If the protection of the whole has not always gone far enough to preserve every individual [state], these are exceptions in the life of the whole [system], which have not destroyed the whole, but have been overcome by it.'[16]

For Clausewitz, then, war was a normal aspect of the relations between states. Yet he did not go as far as Hobbes and Rousseau who saw war as the normal state of affairs in anarchic inter-state relations. For Clausewitz, war could and did frequently result from the conflicting interests of and rivalry between the European states, but it was not the only possible result, nor did it mean that every war was fought with the purpose of the annihilation of states on either side. Clausewitz the Idealist, writing only about the wars of the French Revolution and Napoleon, knew of the *levée en masse* of 1793, the total mobilisation of the population on one side in its efforts to defend itself against its attackers. And having been born in Burg, near Magdeburg, the population of which had been massacred during the Thirty Years War (as Clausewitz noted in his study of this war),

Clausewitz knew of wars which had victimised the civilian population. Nevertheless, while pleading for the *Landwehr* and the people's war to liberate Prussia and other German-speaking lands from French occupation, Clausewitz never included the targeting of the adversary's civilian population as a legitimate war aim (a point to which we will return in Chapter 5).

Clausewitz's amoral discussion of war reflects not only the political reality of his times, but also the status of international law: the very definition of state sovereignty centred on the right to use force, to declare war and make peace, and before the First World War no international convention outlawed war as a legitimate means of furthering a state's political interests. Medieval Christian notions of peace as the desirable state of affairs between Christian princes, which could only exceptionally be broken to redress a just grievance, had long disappeared, and only in 1928 did signatories to the Kellogg Pact express their agreement that war should cease to be a legitimate instrument of inter-state relations.[17]

Clausewitz's Ideas of Society

Clausewitz thus saw war as a normal social activity,[18] but this begs the question of what his ideals of society were. The answer is difficult to find in his main work. Clausewitz's own political ideals are hardly reflected in *On War*, in which he sought to explain warfare in thoroughly apolitical terms. As Peter Paret has rightly noted, 'The general applicability of Clausewitz's theories transcends particular political ideologies.'[19] A level of thinking entirely missing from *On War* is that of ethics and morality.[20] As a disciple of Machiavelli, Clausewitz simply accepted that states are aggressive, and that if some states are aggressive, others have to defend themselves. For Clausewitz, war was a normal part of international affairs, just like other acts of diplomacy and commerce. When he wrote about 'moral issues', he meant the morale of the army and the population of the country involved in the war; indeed, for him, all depended on the degree to which populations were involved, and supported the war effort. The mobilisation of civilians in the *Landwehr* or *Landschutz* was the only possible response to the French onslaught, and Clausewitz therefore saw this as the only way to ward off an enemy who had unlimited war aims. If he unleashed unlimited force against you, drawing on the entire mass of his population, the only defence was to respond in kind, as Austria and Prussia did to some extent, or to find other ways of mobilising the entire population against the enemy, as the Spaniards did in their *guerrilla*.[21] Thus Clausewitz wrote with grudging admiration for the French way of war, taking it as a model of war in its very essence, ideal war.

Occasionally, there are glimmers of sympathy with those who want to avoid unnecessary bloodshed, although the harsh realities of the world seem to Clausewitz to make such an attitude even more dangerous than one which takes war as a necessary evil.[22] The dangers of decadence and the vulnerability of civilised nations confronted with tough barbarians was known to everybody since the publication of Gibbons's *Decline and Fall*, and in *On War* we see echoes of that attitude. Nevertheless, elsewhere Clausewitz described what he called civilised behaviour, and described it with approval:

> If . . . civilised nations do not put their prisoners to death or devastate towns and country, it is because cleverness plays a larger part in their conduct of war and has taught them more effective ways of using force than such a crude expression of instinct.[23]

Elsewhere he raised the question of where military genius is most likely to be found. In barbarian peoples, if they are warlike, the military spirit is spread evenly throughout the entire population. Nevertheless, they produce few real military geniuses. It is when a people is both civilised and war-like that military brilliance is most readily found. As examples, he cited the Romans and, in his day, the French.[24] For Clausewitz, then, the most successful nation is both civilised *and* bellicose. This is not necessarily because he saw a war-like spirit as positive in itself, but we must remember that, faced with Napoleon, Clausewitz's Prussia had not been a bellicose nation; it had fought according to the cautious rules of the eighteenth century, with a professional, not conscript, army, and was duly defeated by France. Clausewitz wrote in the light of these facts, with a sense of bitterness:

> Today practically no means other than war will educate a people in this spirit [of boldness], and it has to be a war waged under daring leadership. Nothing else will counteract the softness of mentality, this yearning for comfort, which debases a people in times of growing prosperity and increasing trade.
>   Only if the character of a people and the habit of war are continually mutually supportive, can a people hope to have a firm position in the political world [i.e., in international relations].[25]

Taken in the context of the rise of nationalism in Germany, particularly as the nineteenth century moved on, this passage of course would lend itself to Darwinistic interpretations, as justification for the militarisation of society, the glorification of war and the apotheosis of the military. But would Clausewitz have seen that as flowing logically from his views, which were the product of the bitter defeat and humiliation of his country?

For the purpose of defeating the French, Clausewitz advocated the raising of a people's territorial army, the *Landwehr*, and in his writing about this subject showed himself well aware of the potential for social revolution which the mobilisation of the masses might bring. It transpires from his writing, however, that he hoped the popular fervour could be guided into purely patriotic channels, allowing the conservation of the monarchical-hierarchical structure of society in which he himself, as a member of the military elite, had a vested interest. He thought that the defensive mobilisation of the civilian population would be particularly effective if the population was suitably accustomed to 'civilian obedience', humbly accepting their position in society as that of a prince's loyal subjects.[26] He wanted the Prussian king's government to surround itself with representatives of the people, not ones chosen by general manhood suffrage as in the French Republic, but similar to the contemporary oligarchic constitution of the British Parliament. Such a structure would enable the government to fight the enemy without and to protect itself against insolent, insurrectionist forces within.[27] Clausewitz himself, as a counter-revolutionary, was realistic enough to see that in the revolutionary Europe of 1789 to the mid-nineteenth century, 'Only such princes could survive who can enter into the true spirit of the great reform, and indeed themselves assume the leadership of it.'[28]

The Trinity: Violence, Chance, Political Purpose

Napoleon said in one of his (otherwise, on the whole, rather banal) 'Maxims': 'Experience proves that armies are not always sufficient to save a Nation; while a Nation defended by its people is ever invincible.'[29]

Clausewitz, from his earliest military experiences, had witnessed the encounter of his own side, a classical army of the *ancien régime*, with the profoundly different, ideologically motivated people's army of France. Civilians had largely been spared the effects of war in the eighteenth century – more so than in previous or later centuries. There were exceptions which Clausewitz does not seem to have been aware of: in the American war of independence, civilian combatants were engaged in great numbers, for example, and the repression of the Jacobite Rebellion (1745–6) subsequent to the battle of Culloden was far-reaching.[30] In general, however, the campaigns of Prussia in the Frederician age had done little harm to the civilian populations.

In his first schooling under Scharnhorst, the reason for the deep transformation of war in the 1790s was explained to Clausewitz. In

Clausewitz's words, 'The enormous outward effects of the French Revolution are apparently to be sought less in the new means and ideas about its conduct of war than in the totally transformed art of statecraft and administration, in the character of the government, in the state of the people.'[31]

From this, he formulated his often-quoted theory of the 'remarkable trinity' that dominates war, composed of:

- the primordial violence of its nature, the hatred and enmity which must be regarded as a blind natural instinct;
- the interplay of probability and chance, which make [war] a free activity of the soul;
- and the subordinate nature of a political tool, through which [war] becomes subject to reason.

> The first of these three sides is turned more towards the people, the second towards the military commander and his army, the third towards the government. The passions which are supposed to ignite in war have to exist within the peoples; the dimensions which the interplay of courage and talent will have in the realm of the probability of chance depend on the peculiarities of the military commander and the army; the political purpose are subject to the government alone.

Theory, he concluded, has to take into account the interplay of all three dimensions.[32]

Raymond Aron noted that Clausewitz applied this trinitarian analysis only to 'absolute' wars, and did not fully explore the effect on a war if the people were not involved, even though it is implicit in his writing;[33] nor did he fully explore the role of his 'trinity' in a people's war, when there is little or no distinction between population and fighters. Nevertheless, praised by some as 'the master concept in Clausewitz's theory of war', this first 'remarkable trinity' – of 'primordial violence, hatred and enmity', 'chance and probability' and the 'element of subordination, as an instrument of policy' has inspired many subsequent strategists.[34]

But it is Clausewitz's secondary trinity – of people, military and government – that has attracted more attention. Mao Zedong was inspired by it to explore the dynamics of the relationship between combatants and the population at large. For Mao's concept of the great revolutionary war, the mobilisation of the people for the purposes of the revolution was essential: the people had to be turned into 'a gigantic ocean, in which the enemy will drown'. In order to succeed, one had to 'link political mobilisation with the development of the war, with the life of the soldiers and the popular masses and turn them into a lasting movement'.[35] Mao thus filled the void in Clausewitz's own teaching about the mobilisation of the masses: in his dislike of the French and of the democratic, republican ideals of the French Revolution, Clausewitz had always avoided putting his finger on the force of the French Revolutionary ideology, discreetly referring to the 'transformed art of statecraft and administration, the character of government and the state of the people'.[36] Mao, by contrast, identified the force of ideology as the main motivation of the masses who had to be mobilised.

Clausewitz's secondary trinity of people, military and government has been widely popularised – to the point where it has almost eclipsed Clausewitz's primary trinity – in a book by Harry G. Summers Jr. that analysed the mistakes made by America in the Vietnam War on the basis of Clausewitzian tenets. Summers emphasised in particular that in a democracy, war is impossible to sustain for any length of time if the people, and all parts of the government, do not identify strongly with a (clearly defined!) war aim and with the armed forces who are sent to achieve it.[37]

Differentiating between people, military and government arguably presupposes the existence of states with governments, populations and armed forces which are not identical with the 'people in arms', and Martin van Creveld and John Keegan have found this wanting as an analytical approach to civil wars or other forms of sub-state conflicts. Moreover it presupposes a state in which there is no conscription but a professional army (so that people and army are not one). It applies neither to the many different social entities – from the Greek and Italian city states, through Roman imperialism to the Holy Roman Empire – that existed before the 'modern' world (marked by the Westphalian Treaty of 1648 with its *de facto* recognition of state sovereignty), nor to the many forms of 'irregular' warfare – regular forces pitched against irregulars, or tribal or gang warfare – which have existed throughout history and which have produced most casualties since 1945. The very concept of the 'state' was alien to most cultures outside Europe until their colonisation by European powers in the nineteenth century and their political independence in the course of the twentieth. Thus van Creveld argues that Clausewitz's trinity does not apply to the majority of periods and wars in history. Van Creveld and Keegan argue that 'trinitarian' warfare is merely one of many forms of warfare, and a very special one at that.[38]

'Non-trinitarian' wars are usually subsumed under the heading 'Low Intensity Warfare', but in many cases – the genocide in Rwanda in the mid-1990s, for example – have produced numbers of casualties comparable to those of 'High Intensity Warfare', particularly among the civilian populations. They generally occur in areas where the state is underdeveloped – in Third World countries – or challenged – as in Northern Ireland or in Turkey. The existence of a state that is fully in control of military operations, which Clausewitz's formula postulates, is thus a phenomenon that is limited in time and place. And although it is likely to continue into the twenty-first century for many Western states, they are likely to find themselves, according to van Creveld's prediction, confronted increasingly with sub-state or simply non-state enemies – from terrorists within their states to parties to a war outside it. The terrorist attack on the US on 11 September 2001 confirmed van Creveld's prophetic writing.

This leads van Creveld to point to a very important weakness in Clausewitz's work: namely the paucity of his comments on the involvement of the people in war. Clausewitz saw them as a major force in the French Revolution and its wars, and as supporters of the combatants, but he did not reflect on their role as victims. In the age of aerial bombardment, however, little distinction between combatants and non-combatants is possible, and Clausewitz's world in which a war was decided on the battlefield has given way to wars that turn whole countries into battlefields. The art of using battles for the objective of the war (Clausewitz's definition of strategy), as van Creveld notes, presupposes the limited reach of weapons and the existence of distinct battlefields, distinctive campaigns, skirmishes that can be identified as such, bases, objectives, distinctive lines of communication, all or several of which may be absent in non-trinitarian war, be it of the past or of the future. Van Creveld notes also that the treatment of prisoners of war, and of civilian populations in occupied territories, like many other aspects of war, has undergone repeated transformation over the centuries. It does not help, therefore, that Clausewitz takes for granted certain limitations on war (such as the rules governing the treatment of prisoners and of civilians of his age), failing to address them as variables in its conduct.[39]

Van Creveld cites a number of other Clausewitzian ideas with approval – from the concept of friction to the comparison of the preparations of war and the conduct of war with the work of a sword-smith and a swordsman respectively – but he is sceptical about the clear definition in each war of a war aim.[40] In his view, Clausewitz, despite his emphasis on it, still underestimates the passions aroused by violence, where in cultures not disciplined by a complex governmental decision-making process, war becomes an end in itself.[41] With the disappearance of trinitarian warfare, i.e., warfare between well-ordered states with their government, armies and citizens, classical strategy will disappear, in van Creveld's view. Thus not all wars have clear war aims – indeed, are not war aims often pretexts for people to engage in the wars they *wish* to engage in?[42]

To sum up, Clausewitz's secondary trinity can help explain phenomena such as the lack of large-scale public support for the wars of the *ancien régime*, or for America during the Vietnam War, and the concept could usefully be developed much further, to explain and analyse the dynamics of civil-military relations, and the relationship of government and armed forces, to which we are now turning. But in other respects, the model is too rigid and limiting. However, as Edward Villacres and Christopher Bassford have argued in a pugnacious article, it is not the secondary trinity, but the primary trinity that truly merits our attention, and it is hardly justified to dismiss Clausewitz as outdated because the secondary trinity does not apply to all forms of war.[43]

The Commander-in-Chief and the War Cabinet: Supremacy of Politics or of Military Strategy?

Clausewitz had more wisdom to offer on civil-military relations. As he explained in *On War*:

> At its highest level, the art of war is politics, but, no doubt, politics which fights battles instead of writing notes.
>
> In view of this fact, to leave a great military event, or the plan for one, to a purely military judgement, is an impermissible distinction and even a harmful one; indeed it is an irrational procedure to consult professional soldiers on war plans in order to let them give purely military advice on what the Cabinet should do; but still more absurd is the demand of theorists that the available means of war should be handed over to the General so that he may draw out a purely military plan for the war or for the campaign, in accordance with those means. Moreover, general experience teaches that despite the great [sophistication] of today's warfare, the main features of war have always been decided by the cabinets, i.e., by a political, not a military, bureaucracy . . .
>
> None of the principal plans necessary for war can be made without an insight into the political situation, and when people speak (as they often do) of the pernicious influence of policy on the conduct of war, in reality they are saying something quite different to what they mean. It is not this influence, but the policy itself which should be criticised. If policy is right – that is, if it succeeds in achieving its aim – then it can only act positively on the war . . . and if this influence of policy leads away from the aim, the cause is only to be sought in a mistaken policy.
>
> Only if policy makers [*die Politik*] wrongly believe that certain

military means ... will have effects which in reality are against their nature, it can ... exercise a pernicious influence on [the conduct of] war. Just as somebody who with the right idea says something wrong from time to time when speaking a language which he does not quite command, civilian governments will in such circumstances often order things which go against their own intentions ....

If a war is supposed to correspond entirely to the intentions of civilian governments, and if the policies are supposed to be appropriate to the means of war, only one solution exists if the political decision-maker (statesman) is not the same person as the military commander: to make the highest military commander a member of the Cabinet, so that it can participate in the main moments of his actions.[44]

Clausewitz thus imagined that the civilian government should determine the way in which the military commander-in-chief operates. But in the second edition of *On War* published in 1853, Count Brühl, along with a number of other editorial changes, reworded this last section to read:

... to make the highest military commander a member of the Cabinet, so that he can participate in its most important deliberations and decisions.[45]

This, however, was clearly not Clausewitz's intention, for we read in a letter of 22 December 1827:

The task and right of the art of war with regard to policy [*Politik*] is mainly to prevent policy makers [*Politik*] from demanding things which are against the nature of war, lest policy makers [*Politik*] commit mistakes in the use of the art of war because of their ignorance of the effects of the instrument.[46]

Even in 1815 he had written that it was a misconception to think that military operations could not be conducted well by a civilian government, citing the French Revolutionary regime as evidence.[47] Also, we find in his original Book VIII, '[P]olicy, of course, is nothing in itself; it is simply the defender of all interests [of one state] against other states.'[48]

The change made by Count Brühl reflects the militaristic culture which increasingly characterised Prussia and other German states. Already the Prussia of the 'soldier-king', the father of Frederick II (the Great), had built its strength on its army. But during the middle and second half of the nineteenth century, the military rose even further in social esteem. Clausewitz himself was embarrassingly proud of his own social standing as (just about) nobleman and, above all else, military officer; and in most social contexts, officers of low rank would be given precedence over all civilians present, including figures such as mayors and ministers. Prussian society underwent what Hans-Ulrich Wehler called a 'social militarisation', in which the military commanded ever greater respect, while civilian leaders lost social credit in relation to the military. The martial successes of the Prussian military also created a bond between the rise of the Prussian and then the German state, in which the military occupied a rank well above that of any civilian leaders, bar the king of Prussia and later emperor of Germany and his family, whose male members significantly appeared in public most frequently dressed in military uniforms. Even Bismarck, as the first chancellor of the Second German Empire, found that he was ignored in many places if he did not wear a military uniform.[49] It must be said that the same tendency existed throughout continental Europe. Much like Wehler, his French colleague Raoul Girardet has called France a 'military society'.[50]

In keeping with this spirit of his age, Colonel (later Field Marshal) von Manteuffel commented on Prussian Prince Frederic Charles's draft of a war plan against Switzerland 1856–7:

> Your Royal Highness postulates that politics are the main aim, war the means for achieving this aim. Yes, before deciding on war, politics must decide; on whether and when to end it, suspend it, etc., politics has a say too. But when the sword has been drawn, war . . . steps into the foreground, becomes fully independent, and politics becomes its servant. A war which is merely conducted as a means to reach a specific political aim is never a real war, it is only a more or less important demonstration; such a [war] can never lead to a result, if the general political situation does not happen to be such that it is not necessary to wage a war, but merely to stage a demonstration in order to reach one's aim, that one only wants to occupy a country and can strike down the small opposing forces without fearing real resistance, in short, if a clearly recognised superiority of forces exists [on one's own side]. If war is the order of the day, its nature must be fully recognised, and its only aim can be the annihilation of the enemy, and all its efforts must be directed towards this end . . .[51]

It was only after the Second World War that the Clausewitz scholar Werner Hahlweg found out that the second edition of *On War* contained the forgery quote above. For a century, however, the debate about the relationship between the military leader and the civilian government was prominent in internal German politics.

Civil and Military Command: The Clash between Bismarck and Moltke

One of the most famous examples of this debate concerns the relationship between the Prussian, and then German, chancellor Bismarck and the chief of the military staff, Helmuth von Moltke the Elder. Ironically, Bismarck admitted 'his shame at never having read Clausewitz, and of not knowing anything about him, except that he was a very deserving general',[52] while Moltke, who had studied at the Prussian military academy while Clausewitz was its director, claimed to be a great admirer of his. Moltke only knew *On War* in its second edition, however, and crucially interpreted Clausewitz's views on civil-military relations against the true Clausewitzian sense. He had always regarded practical issues as more important than abstract theories of warfare.

Under the supreme authority of the Prussian king (who from 1871 was German emperor), and sidelining the minister of war, Moltke and Bismarck formed two points of an uneasy triangle of leadership that was troubled by the competition between the two forceful characters.[53] Moltke wrote in his first operational plan for a campaign against Denmark (before the war between Prussia and Denmark of 1864): 'Neither diplomatic negotiations nor political considerations should interrupt the further military progress.'[54] And in a memorandum to King William of 29 January 1871, Moltke wrote:

> I believe that it would be a good thing to settle my relationship with the Federal Chancellor [Bismarck] definitively. Up till now I have considered that the Chief of the General Staff (especially in war) and the Federal Chancellor are two equally warranted [*berechtigte*] and mutually independent agencies under the direct command of Your Royal Majesty, which have the duty of keeping each other reciprocally informed.[55]

Moltke's attitude reflects a structural concern of the military not to be forced to fight with one hand tied behind its back, but to be given ample means to pursue even limited military goals. Clausewitz's distinction between limited and unlimited political aims and limited and unlimited military means used in pursuit of these aims can hardly please military leaders, who would gladly swat a fly with a sledge hammer if that is the easiest thing to do and will result in minimal losses on one's own side. Moltke for one did not quite deny Clausewitz's concept of the 'dual nature of war', but he simplified it by saying that the quantity and quality of the 'available means' should determine the military aims.[56]

Moltke was thus close in his thinking to that of Clausewitz the Idealist.

In an essay on strategy dating from just after the end of the Franco-Prussian War, he wrote:

> Policy uses war to gain its objectives, it has a decisive influence on its beginning and end, in such a way that it retains the prerogative of increasing its demands during the course [of the war] or to make do with lesser successes. Given this uncertainty, strategy can only and at all times direct its efforts towards the highest goal which it can obtain with the means available. In this way it serves policy best in the attainment of [policy's] objectives, but [strategy] is completely independent of [policy] in its action.[57]

From the beginning of the Austro-Prussian War (2 June 1866), Prussian royal instructions were that the General Staff could henceforth issue orders directly to the troops and did not have to clear them first with the War Ministry. These orders applied for the duration of the war.[58] During the Austro-Prussian War, Bismarck intervened a number of times in what the military regarded as their affairs, raising objections to their plans on technical grounds. After the battle of Königgrätz, he participated in military councils, much to the annoyance of Moltke, who was much admired by King William of Prussia. Indeed, the General Staff agreed that political leaders should in future not be given this right to interfere in military matters.[59] In the Franco-Prussian War (1870–71), therefore, Bismarck was no longer invited to meetings between the King and the military leaders. After the battle of Sedan, Bismarck realised that crucial information was being kept from him, and demanded from Moltke that 'I receive continuous information concerning military proceedings and, if this does not seem possible in any other way, [that I receive it] by means of the simultaneous communication of the telegrams which are designed for the Berlin press, the content of which is still in most cases new to me when I read it five days later in the newspapers.'[60] Bismarck wanted Napoleon III to remain as emperor and to make limited territorial concessions in the form of Germanophone Alsace and the fortified city of Metz in Lorraine. But Napoleon III felt unable to make this concession, and the negotiations came to nothing. As the war continued, in September–December 1870 the question arose of what was to become of Paris. Moltke was willing to besiege it but not to incur significant numbers of casualties through taking it by force, as he wanted to save his armed forces for further encounters with the French, who were increasingly fighting a 'people's war' inspired by French nationalism. Bismarck, however, wanted to take Paris by force, which did indeed happen through bombardment in January 1871. Bismarck did not, however, want German troops to occupy Paris, which was what Moltke demanded in order to destroy France's will to resist.

Bismarck prevailed on both counts, but Moltke in turn succeeded in imposing on France not only the cession of Alsace, but also of a larger part of the strategically important area of Francophone Lorraine, including Metz, giving the French a long-term grievance against the German empire.[61]

In the historian Gordon Craig's words:

> In the three wars of the unification period, Bismarck had successfully maintained the principle of the predominance of politics in war-time. It had, however, become progressively more difficult for him to do so; and, what is more significant, his victory had not convinced the soldiers that he was right ... Generations of officers, trained in the General Staff by Moltke, were to accept [Moltke's] ... demand for a line of demarcation between politics and strategy which would free the strategist from civilian interference ... as a doctrine and were to attempt, with disastrous results, to apply it in the First World War.[62]

Indeed, Moltke's influence grew, and with it the tendency towards an ever more comprehensive way of waging war against an enemy's entire society. In his 'War and Politics', Moltke wrote about 'the mutual influence between politics and strategy':

> War is the forceful action of peoples for the purpose of realising and maintaining the policies of the state; it is the extreme means of acting out the will of doing so, and while it lasts, it creates a situation in which international conventions are suspended between the warring factions.[63]

And Moltke quoted with approval Clausewitz's letter to Müffling (see p.57), in which Clausewitz had argued that the military did have the right to decline technically unfeasible missions, adding:

> For the course of war is determined mainly by military considerations, while political considerations are only of avail if they do not ask for things that are military nonsense. In no case must the leader allow his operations to be influenced by politics *alone*; much to the contrary, he must keep his eye on military success. It is of no concern to him how politics can subsequently use his victories or defeats: it is up to politics to exploit them. Where the head of state goes along on a campaign, as is the case with us, the political and military demands are reconciled in one person.[64]

For Moltke and his successors, it was thus preferable that a commander-in-chief should decide on the nature of the war effort rather than allowing a civilian leader to interfere in the conduct of war.

Military Planning and the Lack of Civilian Guidance

On the eve of the Great War, nobody in Europe contested the notion that general decisions on war and peace fell to the civilian leadership, and that the military should have no say in this area. However, on all sides detailed war plans were nevertheless drawn up by the military, leaving political considerations outside the purview of the planning. In Germany, Schlieffen thus drew up what the military historian Jehuda Wallach has called a '"purely military plan" without consulting any of the political agencies, and the result was disaster'.[65] Wallach noted, 'Strange as it may seem, the famous Schlieffen Plan is by no means a comprehensive plan for war, embracing all of its aspects – political, economical and military – but it is only the plan for the German ground forces in case of war.' The plan presupposed that hostilities with France would precede hostilities with Russia, while real events in 1914 turned out to be the inverse.[66] Lack of political guidance was not typical only for Germany. Indeed, as Jehuda Wallach rightly summed up, 'The First World War was on both sides of the front lines void of any strategic ideas.'[67] As Ulrich Marwedel observed between the Franco-Prussian War and the First World War, 'the tendency of enlarging the competences of the military leadership at the cost of the political leadership predominated' particularly in Germany. Oddly, Clausewitz's teaching of war as the continuation of politics by other means was taken as an argument for the *division* of these spheres, with an emphasis on the *otherness* of the means employed. The military leadership used this argument to call for a redistribution of tasks and responsibilities within government, with them as beneficiaries.[68] Thus, during the 1871–1914 period, the army was successful in preventing parliament from controlling its activities. The officer corps continued to be recruited mainly from the nobility and retained what Gordon Craig has called a 'feudal relationship to the Crown'.[69]

This is reflected in the strategic literature of the time. In the preface of the 1880 edition of *On War*, Colonel Wilhelm von Scherff wrote: 'The interference of policy in the conduct of war leads always to ruin. Policy sets the fashion *how* the house should be built, but it must not interfere with the building process itself.'[70] The widely read military author Colmar von der Goltz wrote in 1883: 'It behoves us, therefore, to have a sharp eye for, and to guard against . . . the interference of political considerations with the strategic and tactical decision.'[71] Elsewhere he wrote: 'War will . . . be in no way lowered in importance nor restricted in its independence, if only the commander-in-chief and the leading statesmen are agreed that, under all circumstances, war serves the end of politics best by a complete defeat of the enemy ... This leads us back to the conviction already expressed that a state is most happily situated when commander-in-chief and statesman are combined in the person of a great king.'[72] General Rudolph von Caemmerer, however, sided with Clausewitz against Moltke in the debate. He claimed to be 'convinced ... of the correctness in every respect of Clausewitz's views'. Such outright support for Clausewitz (in Brühl's altered edition) was rare, though. More typical was General Julius von Verdy du Vernois, who in his *Studien über den Krieg* ('Studies on War') thought both Clausewitz and Moltke too extreme, taking a middle course between the two.[73]

Similar questions about military competence troubled the British. In 1908, there was a debate at the British Staff College involving the Chief of the General Staff, General Sir Neville Littleton, Major-General Douglas Haig and Brigadier-General Henry Rawlinson. Sir Henry Wilson, the Commandant of the Staff College, was accused of having encouraged the military students of the college to debate political issues which were outside the scope of their competence. Wilson's teaching methods did not change, however.[74]

French military thinkers were also aware of the dispute about Clausewitzian teaching in Germany. General Iung, writing in 1890, specifically referred to the Prussian views on the relationship between civilian government and military leadership, claiming with approval that the Prussian recipe for victory was for the civilians to designate a war aim and then give the military commanders total freedom in pursuing this aim. Politics, i.e., the government, should give the military commanders 'the initial point for strategy'; 'in the entire possible sequence of [ensuing] events that result from strategy and tactics, politics intervenes one single time, when it chooses the goal.'[75]

From his own proposition that limited war was no longer conceivable in Europe, Colonel, later General, Colin (one of Clausewitz's French disciples) deducted that 'the instructions which a government has to give to the general with regard to the political objective of the war are thus reduced to very little', namely, to achieve a decisive victory in decisive engagements. In 1911 he wrote,

Once the war is decided upon, it is absolutely necessary that the general should remain free to conduct it in the way he sees fit, unless he is relieved of his command if he exercises it with little energy or competence.

Indeed, Colin argued that 'The campaign plan has to be the personal work of the general . . . It has almost never happened that the intervention of a government in the conduct of operations has produced happy results.'[76]

Before the great cataclysm, the chief of the historical section of the Grand General Staff of the German Empire, General Friedrich von Bernhardi wrote,

> War is only a means of reaching a purpose that lies well outside its sphere. War can therefore not define the purpose for itself by determining the military aim according to its own considerations. If one wanted to concede to war this prerogative, there would always be the danger that the war, stripped of all constraints, would be waged for its own sake, or else that its achievements would fall behind the politically necessary targets . . .
>
> If [politics] defines its purposes, taking into account the tools of power available to the state, and in times of war determines the military aims in co-operation with the military commander-in-chief, it must *never become involved in the conduct of war itself* or *seek to dictate the ways* in which the military aim is to be reached . . . War is a continuation of politics with *other* means and becomes a traitor to its very essence as soon as it uses political means . . .
>
> If policy-makers wanted to steer military actions into political channels, and use war not as an independent tool of politics, but as a political tool, military success would be unlikely. Even the very attempt to influence military actions through political considerations has repeatedly led to the most awful consequences.

For Bernhardi the military commander-in-chief had to have a particularly strong character in order to withstand demands on him coming from the political leadership, which he would have to reject on military grounds.[77]

World War I: The War of the Generals

The French Prime Minister Georges Clemenceau famously remarked in 1918 that 'war is too serious a business to be left to the generals'. But this is precisely what political leaders, particularly in Germany, did, and responsibility for this aberration was wrongly laid at the door of Clausewitz. After the First World War, Field Marshal Paul von Hindenburg wrote in his memoirs:

> There is a book 'On War' which never becomes obsolete. Its author is Clausewitz. He knew war and he knew human beings. We had to listen

to him, and if we followed him it would be to our advantage. The reverse meant disaster [*Unheil*]. He warned against encroachments of politics upon the conduct of war [*sic*].[78]

After the war, an eyewitness testified before a Reichstag Committee of Inquiry: '[General] Ludendorff considered the conduct of war, and politics, to be two separate entities, fighting each other with hostility.'[79]

Jehuda Wallach shows that the German Chancellor Bethmann Hollweg's memoirs indicate that he 'unconditionally surrendered his legitimate position as the leading statesman and consciously yielded to the dictates of the General Staff'.[80] Bethmann Hollweg claimed:

Just as with the initiation of the war, political measures had to be fashioned in accordance with the needs of the campaign plan – declared to be unalterable –, so too, during the war, the military view of technical possibilities and strategic effects decisively determined the great operations. The political leadership did not participate in the drafting of the campaign plan, nor in the alterations which Schlieffen's plan underwent from some time before the outbreak of war, nor in the changes made by the modified plan in its application. Generally, during my term in office, nothing like a war council ever met, in which policy might have interfered with the military deliberations.

The only exception had been the decision to launch unlimited submarine warfare, as the political consideration that this was likely to bring the USA into the war weighed heavily, but of course this political objection was overruled. Bethmann Hollweg himself thought that there was no way

the military layman could claim the right to judge military possibilities, let alone military needs. My recollection is that it was military needs that guided the conduct of war. Even the most brilliant initiative undertaken by the General Staff was influenced by great constraints. It was possible only for the military to decide how to cope with these constraints, even in cases where both military and political demands existed.[81]

After the First World War, in his book *The War of the Future*, Friedrich von Bernhardi drew lessons which were diametrically opposed to those drawn by post-World War II strategists such as Jehuda Wallach and Bernard Brodie. Bernhardi relegated politics even more firmly to the back seat, postulating that

the task of diplomacy is quite different when war has actually broken out. It still has the duty of keeping other states from participating in the war ... But this ... must only be [done] in agreement with the Supreme Command. In all other respects the only duty of diplomacy is to support the military with all its might. It must conform to their wishes entirely, and give up any idea of taking any steps without consulting them. That is equally necessary, of course, in the case of affairs which may possibly lead to war – for there must always be a proper relation between the statesman and the general. But a breach of this rule never produces its effects more directly and immediately than in war itself. The military penalty follows the political mistake at once. Diplomacy must therefore confine itself to preparing the way for military victories and exploiting them, but only in accordance with instructions *to be given by the military authorities.* Where this rule is not observed, military and political measures might pursue the same aim but in a totally different spirit, and that might easily yield wholly contradictory results. The upshot is that, where possible, military and political directions should be combined in one person ...

No one has ever enjoyed quite the same position as [Frederick the Great], and where that is so the political authority ... must know how to keep in the background and leave the military authority to decide what is to be done. The politician must unconditionally submit to the will of the soldier, for the combination of political and military action is the main object, and military requirements determine the political. The military leader must be selected accordingly, and where that cannot be done it is much better that someone less expert in politics should settle the broad lines of diplomacy than that the soldier and the statesman should work against each other.

As long as the war is in progress and there is no immediate prospect of a suitable peace, military victory alone must be pursued and everything else must be directed to furthering those efforts. If peace is in sight, on the other hand, it is again for the soldier to judge whether it is necessary to secure it by intensifying the military effort or whether it is better to adopt the diplomatic method, i.e., by making concessions. Only the soldier is in a position to judge. Not the least of the causes of Germany's downfall [in 1918] is the fact that this simple rule was not observed.[82]

General Erich Ludendorff (1865–1937), who virtually assumed the political as well as the military leadership of Germany towards the end of the First World War, wrote:

The time had gone when policy-makers [*die Politik*] could tell the military leadership [*Kriegsführung*], 'win the war, the rest is my job' ...

Military leadership and politics had become one. There could no longer be any doubt that the integrated policy of the state had to serve the war and to fulfil its needs ... The problem of peace was seen by the [Highest Military Command] as falling into the competence of policy-makers [*der Politik*]. In that the [Highest Military Command] stood on Clausewitzian ground. Nevertheless, without jeopardising the conclusion of peace or prolonging the war, it had to ensure that the peace policy of the [imperial chancellor] did not harm the conduct of war, and to work towards a definition of frontiers, which, building on the experience of the war, would provide favourable military and economic conditions for the successful conduct of a new war. In addition, in the case of peace agreements concluded before the war had been brought to an end everywhere, one had to ensure that immediate disadvantages should not accrue to the conduct of war from the negotiations or conditions.[83]

In his book on 'Total War', published in 1935, Ludendorff advocated that the nation should choose a leading military figure to govern it in times of peace, to prepare for war, and then lead it into the war, as only a military leader would understand the needs of war and initiate the right preparations in peacetime. Here he finally turned the Clausewitzian concept completely on its head, postulating that all politics should be dominated by war.[84]

Ludendorff was not the only German dreaming of renewing the war and of setting up a totalitarian state – he was in the company of Hitler, of course, with whom he founded the National Socialist Party, but also of the sociologist Hans Freyer, the writer Ernst Jünger, the lawyer Ernst Forsthoff and many others, who in their apotheosis of the state saw in war the best means of consolidating the state's absolute dominance of society. It is not accidental that it should have been a university professor like Freyer in Germany and a Soviet military writer like Shaposhnikov who respectively described politics and peace as 'the continuation of war with other means'.[85] For Freyer, the state could only unfold its power entirely if it was at war; for Jünger, the mobilisation of the entire population was highly desirable and could be brought about in the context of war; for Forsthoff, the state had to be racially purified by destroying all inimical elements within it, among whom he listed particularly the Jews.[86]

General Hans von Seeckt (1866–1936), the creator of the *Reichswehr* of the Weimar Republic, quoted Clausewitz approvingly with regard to the primacy of political considerations over military execution, although he thought that 'obviously the choice of the way to get to the aim must be left to the military commander-in-chief, even though political considerations will intrude on this choice and that of the means.' In his view Clausewitz demanded 'permeation of the conduct of war by the ideas of the state which makes use of many forces and means, and to which all are subordinate'. It was the state which had to assure the match of political purpose and war aim.[87] Differently put, Seeckt called 'war the bankruptcy of politics'.[88] In practice, however, the *Reichswehr* made little effort, prior to 1933, to accommodate political orders in their military planning. The reason becomes clear when we read some other *dicta* of von Seeckt about the army in the Weimar Republic he so despised: '"Hands off the army" I call out to all political parties. The Army serves the State, only the State, because it is the State.'[89] But he wrote also on the relationship between politics and war:

> The basic situation at the beginning must relate to the political situation. If the conduct of war itself and the ability to wage war are simply the most important elements in the politics of a great State, politics in turn continually influences the conduct of war. It is a great error of political and military-political, but also of strategic doctrine to claim that the conduct of war takes place in a . . . political vacuum, and that it could one day be locked away again, yielding the stage to pure politics. The difference in the manifestations on the one hand of exchanges of notes, economic pressure, ultimata and on the other hand actual hot war is emphasised excessively, while in reality politics and war are of the same nature, striving for dominance on the way to the annihilation of the enemy. Every healthy policy chooses this way which is recognised as the only right one for a healthy strategy. I call both healthy if they conform to their nature, which is after all founded on the struggle for survival, i.e., the fight for dominance. And fighting is to aim for the annihilation of the adversary; peace is the short-lived enjoyment of power.[90]

The 'annihilation of the inimical forces' was the obvious aim of any war for von Seeckt, and he thought the chief-of-staff needed 'full freedom' for the execution of this task.[91]

Somewhat more moderately the lecturer at the *Kriegsakademie* General Wilhelm von Blume, had written just before the Great War:

> War and politics, particularly foreign policy, have a close relationship with each other, resulting directly from the purpose and essence of both. The task of foreign policy is to preserve and further the State's external interests. War is its final, emphatic means, the canon is the *ultima ratio regis*. From this follows initially that foreign policy is very greatly influenced, if not conditioned, by the military capabilities of the State . . .

As war is a means of politics or, as Clausewitz described the relationship so accurately, the continuation of politics with other means ... politics should have the decisive word in the decision on war and peace. Politics cannot, however, make do without military judgement. The knowledge and the right appreciation of the military conditions are the main basis for the decision to go to war, for the determination of the political aims that are to be pursued by war, and for the decision to enter into peace negotiations, and for the negotiations themselves ... Nevertheless it is impossible to direct military actions to the full satisfaction of all expectations in the absence of political insight, without judgement of the political situation and without the recognition of the advantages which it offers.[92]

Ludendorff's successor in the Supreme Command of the Armed Forces towards the end of the First World War, General Wilhelm Groener, was less extreme than his predecessor. Groener wrote in 1930:

The frequently-advocated view that the statesman has no say in the operations, but ought simply to wait until the commander-in-chief reports either victory or defeat or neither, can be seen as obsolete after the experience of the World War. It would, however, be completely inappropriate if the statesman should so to speak claim the role of a strategic supreme controller. Yet one should concede to him the right, or rather obligation, to verify himself whether the presumed results of an operation harmonise with the intentions and aims of policy.[93]

As we have seen Blume deviated from Clausewitz in postulating a high degree of equality between military and civilian command, and Blume's views were shared by Hitler's first Chief of the General Staff, General Ludwig Beck, who wrote: 'The political purpose of the war must be obvious and it must also include in its calculation the final act of every war, the achievement of peace. Only if there is a clearly defined purpose is it possible to deduce from it, and from the available means, the objective of a campaign.'[94] Beck, like Blume, stipulated that the military commander and the civilian commander should rank equally,[95] and demanded 'the right of the supreme military commander to participate in controlling and directing the domestic administration, the organisation of the public food supply, the war economy (finances included), and in controlling and in sustaining the morale of the nation'.[96] Beck went far enough in Clausewitz's direction to disagree strongly and outspokenly with Ludendorff's insistence on the superiority of the military command, but he also criticised the Supreme Commander of the Allied Forces in the First World War, the French

Marshal Foch, for having failed to take part in the deliberations of the Paris peace settlement after the First World War.[97]

The next generation of military analysts drew quite different lessons from the two World Wars. Bernard Brodie pointed out that the military leaders on all sides of the First World War had aimed for military victory without thinking about the political purpose of such a victory, and on the whole had looked down upon their political leaders as ignorant of the military preconditions for such a victory and all too concerned with winning their next election. Civilians had been just as obsessed as the military leadership with the pursuit of military victory for its own sake.[98] Hitler also kept military and political matters strictly separate, as Jehuda Wallach argued.[99] Yet looking at the relationship between the *Reichswehr*, later renamed *Wehrmacht*, and the National Socialist regime, it is striking that, in contrast to the situation during the First World War, the military was all too unquestioningly obedient to Hitler and his henchmen.

In the Second World War, Bernard Brodie argued, the predominance of politicians over military leaders did not guarantee the predominance of political purpose over the military aim of 'fighting to win'. 'The supremacy in authority of the civilian leader', wrote Brodie, 'merely favours the possibility, but certainly does not guarantee, that political purpose will dominate strategy.' Political leaders were all too often convinced by the military leaders that military requirements – 'the exigencies of war' – had to come first. And military leaders live more comfortably with a substantial margin of superiority over the enemy, and with the aim of a clear-cut – 'absolute' – victory. This is how Brodie explained the espousal of the war aim of 'unconditional surrender' of the enemy by Roosevelt at the end of World War II.[100]

In his *War and Politics*, Brodie gave his interpretation of European and American military policy from World War I to Vietnam on the basis of Clausewitz's paradigm of the 'intimate and pervasive connections between politics and strategy', highlighting amongst other things the relations between civilian government and military leadership. He quoted at length Clausewitz's views on the need for military advice in the formulation of war plans, and the need for understanding what the military can do on the part of governing politicians, but added also Clausewitz's *caveat* about the need to subordinate the military point of view to the political one.[101] Reflecting the same scepticism of military war aims that Brodie and Wallach showed, Mao Zedong cautioned: 'Should the military, in a defensive war, develop a tendency to underestimate politics and to try to make the conduct of war something absolute, this would be wrong and must be corrected.'[102]

*

Thus Clausewitz's views on civil-military relations and decision-making – in both their original and altered form – sparked off almost unending debates. The militaristic world of Europe in the second half of the nineteenth century could not accept the idea of civilian policy-makers dominating the military. Consequently, World War I was not fought in pursuit of clear political aims, but in pursuit of victory for its own sake, and turned into a bloodbath beyond Clausewitz's darkest imagination. In the second half of the twentieth century, particularly in the American way of war, we find almost the opposite extreme, namely the political micro-management of war, as in the case of Vietnam (and see the discussion in Chapter 7). There will always be a tension between what military leaders perceive as necessary conditions for a successful military campaign, and what politicians are prepared to grant. Clausewitz's solution to the problem, the presence of the military commander-in-chief when policy-makers determine the political aims of a war which he then translates into military war aims, was an intelligent compromise that, had it been applied on all sides in the Great War, might have avoided the most pointless butchery of armed forces that Europe has seen. It was in rejecting or misinterpreting the Clausewitzian compromise that military leaders opened the way for the hecatombes of the Somme and Verdun, the purpose of which Clausewitz would not have understood.

On the other end of the spectrum, there have also been wars since Clausewitz's time in which political decision-makers failed to put the military instrument to truly appropriate use – and where the military leadership would have done well to point out the discrepancies between political aim and military capabilities. In analysing the latter, Clausewitz made further important discoveries to which we shall turn in the following.

4

Beyond Numbers: Genius, Morale, Concentration of Forces, Will and Friction

Clausewitz's reflections on the political dimension of war, for which he is most famous today, were not what originally gained him most popularity among military thinkers. It was the more practical considerations that he introduced into writing about war which made it possible to explain the fortunes of war beyond thinking in terms of mere numbers. There were several Clausewitzian themes which found their echoes in later literature, as we shall see presently. One concerned the importance of military genius, a second the concentration of forces on the decisive point and the decisive battle, and a third will-power and morale. A fourth was economy of force, and a fifth, which has met with much praise from those who have experienced war at first hand, was that of friction and chance. While, again, Clausewitz derived most of his insights from his close study of Napoleonic warfare and thus did not 'invent' the concepts, his articulation of them inspired many other authors and has indirectly influenced writing on military doctrine around the world.

Genius: The Personality of the Commander

Clausewitz was not of course the first to draw attention to the importance of the commander-in-chief, but he sought a deeper understanding of what constituted military genius. In seeking to articulate the special abilities of particularly gifted military commanders, Clausewitz borrowed from Immanuel Kant, who had written, 'Genius is the talent (natural endowment) which gives the rule to art . . . [it] is a talent for producing that for which no definite rule can be given.'[1] Nevertheless, as Thomas Otte has rightly pointed out, when trying to define military genius,

Clausewitz was less concerned with gut feelings than with the workings of the 'superior intellect'.[2]

Statistically, as we have seen, Clausewitz expected to find military genius more frequently in peoples who combined sophistication with a warlike spirit, for example, the Romans and the French. Clausewitz emphasises that such genius, although arising in this context, was not something that could be learnt systematically: more often than not it is the mark of genius that it transcends rules. Moreover, in war, decisions have to be made very quickly, with a *coup d'oeil*, an intuitive understanding of a complex situation. Clausewitz approved of Napoleon's tenet that the decisions with which a commander-in-chief was confronted amounted to mathematical calculations requiring the genius of a Newton or an Euler; obviously, only an instinctive reaction made with the intuition of military genius could replace this. But intuition was not enough: in addition, a good commander-in-chief has to be steadfast, courageous, energetic and self-disciplined.[3]

Clausewitz's notion that the military genius is to a large extent a natural gift was widely shared among the Prussian reformers. Berenhorst wrote, 'Someone born to be supreme military commander [*Feldherr*] is son of destiny and nature . . . His name is already supreme military commander at his birth.'[4] F. Constantin von Lossau thought that 'To study strategy means to appropriate the characteristics [necessary to command an army], in as far as this is possible, for genius and talents cannot be forced into existence through studies.'[5] And elsewhere he wrote,

> The preponderant and eminent genius, which after a series of adversities finally seizes the opportune moment and finally defeats the inferior intelligence cannot be cast into a scientific formula in order to be taught. And this [genius] is precisely that highest [power] which influences the fate of the states in the place of the god of war.[6]

Somewhat unjustly, the contemporary political scientist Stephen Cimbala has accused Clausewitz of not having developed the concept of 'military genius . . . rigorously. He fails to offer a convincing explanation of the difference between the genius of individual commanders and the collective competency of armies.'[7] Clausewitz did make it clear, however, that the genius of the military leader is but one of the intangible factors influencing the outcome of wars, along with the morale of the fighting forces.

Clausewitz's treatment of military genius inspired other Prussian writers, among them Friedrich von Bernhardi, Count Alfred von Schlieffen and Hugo von Freytag-Loringhoven.[8] And to this day, aspiring military leaders quote him in their quest to understand the intangible factor of the individual talents of the commander.

Centre of Gravity

The second of Clausewitz's themes is at first sight a very clear-cut, technical prescription for success: analysing Napoleon's campaigns, Clausewitz the Idealist saw the route to victory as the defeat and annihilation of the enemy's army by striking at its 'centre of gravity'.[9] Already in his lessons for the Crown Prince, Clausewitz had written:

> We must select for our attack one point of the enemy's position (i.e., one section of his troops – a division, a corps) and attack it with great superiority, leaving the rest of his army in uncertainty but keeping it occupied. This is the only way that we can use an equal or smaller force to fight with advantage and thus with a chance of success. The weaker we are, the fewer troops we should use to keep the enemy occupied at unimportant points, in order to be as strong as possible at the decisive point . . .
>
> Even though we are strong, we should still direct our main attack against one point only. In that way we shall gain more strength at this point. For to surround an army completely is possible only in rare cases and requires tremendous physical or moral superiority. It is possible, however, to cut off the enemy's line of retreat at one point of his flank and thereby already gain great success.[10]

In Book VI of *On War* he defined this in mechanical terms as the point where one could get most leverage, the place where most armed forces were concentrated. The armed forces were expected to follow two contradictory principles: in order to occupy the enemy's land, they had to disperse over a wide area; in order to defeat his army, they had to concentrate in order to strike at his centre of gravity. Clausewitz the Idealist regarded the latter as more important for the purpose of achieving a victory in the main battle. He strongly opposed the division of one's army into separate forces except in very special conditions, as he regarded the striking with all one's forces at once at the centre of the enemy's gravity as the key to success.[11] Thus the identification of the enemy's centre of gravity was the first task in the conception of a war plan, while the second task would be to concentrate the forces needed to strike at it.[12]

The concept of the centre of gravity flows from Clausewitz's study of Napoleon's war-waging technique. But it is through Clausewitz's interpretation of Napoleon's technique that it has entered the *curricula* of military academies and consequently the heads of military planners. Moltke for one agreed with Clausewitz that 'all strategic effects can be traced back to certain centres of gravity'.[13] In Britain the concept of the 'centre of gravity' impressed as well. Writing during the First World War, the journalist Lieutenant Colonel Charles À Court Repington stressed its importance, along with Clausewitz's concepts of absolute war and superiority in numbers.[14] And Major General Sir Frederick Maurice, writing in 1929, showed the commonality between the principles of the British Army's Field Service Regulations and the chapter headings of Clausewitz's Book III of *On War*. The principles were concentration, economy of force, surprise, mobility, offensive action, co-operation and security. Of these, he thought Clausewitz – whom he called 'the father of modern strategical study' – stressed five, including of course 'concentration'.[15]

The concept experienced a great revival in the US when army planners were reflecting on the defeat in Vietnam: there, US strategy had been *not* to strike at the enemy's centre of gravity, nor to move quickly from battle to battle, but to execute very limited strikes, leaving Vietcong morale unbeaten and leaving most of North Vietnam a sanctuary from which the Vietcong could recruit ever more reinforcements.

As a result, and with this Clausewitzian concept in mind, in the 1970s the US invented their new 'doctrine' of the air-land battle, which aimed to intercept and destroy the second and subsequent waves of oncoming enemy forces. The concept of deep strikes against the enemy's forces, limited only in that they might not be nuclear, and the concepts of utmost mobility, forward movements, encircling movements, etc., can all be seen as inspired by Clausewitz.

The American strategist Michael Handel commented on the performance of the coalition forces in the Gulf War of 1990–91: 'The Gulf War was terminated prematurely because its center of gravity was incorrectly identified (according to traditional military terms) as the Iraqi armed forces, whereas the defeat of these forces was in fact only a preliminary condition for attacking the true political center of gravity: Saddam Hussein himself.'[16] Did Handel exceed Clausewitz's definitions? Again we find that Clausewitz transcended the limits of his own earlier definition. In pondering the importance of political factors in Book VIII, Clausewitz the Realist concluded that a defeat of the enemy's armed forces in battle might leave the morale of the enemy only slightly bruised if he had popular enthusiasm to support him: if a nation of 30 million was the enemy, then the defeat of a single army would not constitute a strike at the true centre of gravity. That, he decided, could also be the capital city of the enemy nation (for example, the capture of Vienna by Napoleon after the battle of Ulm in 1805, and the capture of Berlin after the battles of Jena and Auerstedt in 1806). Furthermore, a key aspect of the Napoleonic (and hence, of Clausewitz's ideal) war, was the involvement of the population itself in warfare. Therefore Clausewitz also recognised that a vital war aim was 'to win over public opinion'. The Napoleonic Wars had shown, wrote Clausewitz, 'what enormous factor is the heart and the mind of the nation in the product of forces of state, of war and armed forces. Since the governments have all got to know these auxiliary means, it is not to be expected that in future wars these will not be drawn upon if the existence of the governments is threatened or they are driven by strong ambitions.'[17] But he did not propose to fight for the hearts and minds of the people through ideological appeal and persuasion: crude force – namely big, spectacular victories – and the conquest of the capital were his recipes for success.[18] In this context, Clausewitz advocated the *Volkskrieg*, the people's war: arming the entire able-bodied male population, confident that he could build on their patriotism and monarchism rather than on any promise of redistribution of wealth and power.

In *On War*, we thus find *in embryo* the various elements of a concept which, beyond targeting the enemy army, aimed at changing the will of the enemy more generally.

Thus in the writing of Clausewitz the Realist, the centre of gravity could be many different things – the enemy's army, capital, public opinion. In the case of a small enemy fighting alongside a stronger enemy, the centre of gravity might not be the small enemy at all, but his larger ally. It might be the leader of a resistance movement, or public opinion.[19] Indeed, towards the end of his life, Clausewitz also advocated that one should strike against the main leader of a rebellious group, as such a popular leader, too, constituted the enemy's centre of gravity.[20] We therefore see that, as Clausewitz pondered more on the role of intangible factors like morale, he moved away from analysing war as merely a function of physical factors and military balances.

Concentration of Forces

If the centre of gravity is taken to mean the enemy's armed forces, then to hit the enemy in his centre of gravity requires a particular use of one's own forces, the concentration of one's effort upon this most sensitive point. Again, Clausewitz's insight can be traced back to Napoleon, who mused:

> Gustavus Adolphus, Turenne and Frederick, as also Alexander, Hannibal and Caesar have all acted on the same principles. To keep your forces united, to be vulnerable at no point, to bear down with rapidity upon important points – these are the principles that ensure victory.[21]

Napoleon's knack for concentrating his forces for attack was recognised by Clausewitz's teacher, Scharnhorst, who wrote, 'If possible, one had to direct one's main force upon one point and make this the main point of the entire [operation].'[22] Indeed, Scharnhorst and, following him, Clausewitz postulated: 'Never stand [still] concentrating [your forces] – but always fight concentrating [them].'[23]

Not surprisingly, Napoleon's analyst Jomini had the same recipe for success, namely 'the employment of masses upon the decisive points'; he also wrote about 'decisive strategic points' and 'objective points', and the 'strike in the most decisive direction'.[24] But Jomini put more emphasis on Napoleon's *manoeuvre sur les derrières*, the encircling approach, which Clausewitz did not put at the centre of his reflections, and for which Colonel Hubert Camon, a French military theorist, criticised him.[25]

Even the greatest strategist borders on the banal, however, when he writes: 'The best strategy is always to be very strong, first in general, and then at the decisive point', yet Clausewitz felt it necessary to spell out the advice that the commander-in-chief must always keep his forces together, except if very special circumstances demand their separation.[26] Even back in 1810–12 he had advised the Crown Prince to concentrate all his forces in one point, the decisive point, particularly if he did not have many soldiers.[27] And yet Clausewitz's writing about the concentration of forces again proved ambiguous enough to allow some writers to deduce that only superiority in numbers could produce success, while others took his views on morale to mean that a qualitatively superior army could well win over a quantitatively superior one.[28]

Moltke echoed Clausewitz's demand for large forces:

In all circumstances we must bring to battle whatever forces we can muster, for one can never have too much strength ready nor too much luck for a victory. In order to achieve it, we have to bring even the last battalion on to the battlefield. Many battles have been decided through troops which only reached the battlefield on the eve of the day of battle . . .

If the concentrated forces of the still dispersed enemy would be larger than ours, we must not wait for him to concentrate his forces, but we must, if there is any chance at all of a partial success, attack with whatever forces we have available.[29]

In France, Captain Gilbert wrote approvingly that Napoleon's strategy, as interpreted by Clausewitz, could be reduced to two ideas: firstly, always to have as strong a fighting force as possible, and secondly, to concentrate it on the decisive point with the aim of annihilating the enemy.[30] Ferdinand Foch, in his *Principles of War* of 1903, claimed the 'old theory' had been that 'in order to conquer, you must have superior numbers, better rifles, better guns, more skilfully chosen positions'. But the French Revolutionary and Napoleonic Wars showed that these preconditions were not crucial as long as one had 'superiority of numbers at the decisive point' and stimulated 'our own spirit to a maximum . . . breaking yours.'[31] The old theories, in Foch's view, had led to poor teaching at the military schools which put excessive emphasis on 'the *material* side of the subject'. He continued:

> 1870 woke us out of that sleep, for it gave us an enemy formed by the teaching of history – by the study of concrete facts. It was in such a fashion that Scharnhorst, Willisen and Clausewitz had, from the beginning of the nineteenth century, formed the Command of the Prussian army.[32]

Foch quoted Clausewitz as saying, 'War is produced by, and receives its form from, the ideas, feelings and relations which obtain at the moment it breaks out.' He continued:

> It is because we ignored that radical transformation among our neighbours and the consequences it was bound to bring about that we, who had created national war, became its victims . . .
>
> It is because the whole of Europe has now come back to the national thesis, and therefore to armed nations, we stand compelled to-day to take up again the *absolute* concept of war, such as it results from history.[33]

Foch further postulated: 'If you want to push the enemy back, *beat him*; otherwise, nothing is done; and there is only one means of doing this: namely fighting. "Blood is the price of victory. You must either resort to it or give up waging war. All reasons of humanity which you might advance will only expose you to being beaten by a less sentimental adversary" (Clausewitz).'[34] Again he quoted Clausewitz:

> Bonaparte always marched straight up to the goal without troubling in the least about the strategical plan of the enemy; knowing that everything depends on tactical results and never doubting that he would get them, he always and everywhere sought an opportunity for battle . . .[35]

This Clausewitzian terminology echoes back from military writing around the world. According to the official Soviet military dictionary, Lenin emphasised the most important principles for conducting armed combat: determining the main danger and the direction of the main attack; concentration of forces and weapons in the decisive place at the decisive moment; securing by all methods and means of struggle their use in accordance with existing conditions; the decisive role of the offensive; the objective evaluation of opposing forces; initiative and surprise; firmness and decisiveness; securing success, maneuver of forces; and pursuit of the enemy to his full destruction.[36]

Even where strategic practice deviates from that advocated by Clausewitz, we hear the echo of his writing. Liddell Hart also criticised Clausewitz's tenet that superior numbers was the decisive factor in achieving victory. (Clausewitz: 'An unbiased examination of modern military history leads to the conviction that superiority in numbers becomes every day more decisive; the principle of assembling the greatest possible numbers may therefore be regarded as more important than ever.'[37]) Liddell Hart pointed to the horrendous results of the misapplication of Clausewitz's teaching at the time of the First World War:

Through Clausewitz [this principle] would become more important than all else in the thought of the European military chiefs. Under his influence, they would neglect to develop the latent superiority that mechanical invention increasingly offered. Only with reluctance would they accept the new tools forced on them by civil progress, causing an immense and needless time lag between their invention and provision. That time lag caused the needless massacre of millions. So did the fact that for a century the soldiers of Europe would cling to Clausewitz's delusion that 'the close combat, man to man, is plainly to be regarded as the real basis of combat'. . . . In reality it meant that they were training their masses to be massacred by machine-guns.[38]

In the First World War, according to Liddell Hart,

the theory of mass suffered [a] rude . . . shock. Calculated to achieve success by a process of concentrating superior numbers at a so-called decisive spot, the formula was nullified by the mechanical progress which made one man sitting behind a machine-gun the superior of a hundred, sometimes a thousand, who were advancing upon him with a bayonet.[39]

The call for a concentration of forces against the enemy's centre of gravity lived on, albeit in a more sophisticated guise. In the US Marine Corps' manual *Warfighting* of 1994 we read:

We have since come to prefer pitting strength against weakness. Applying the term in modern warfare, we must make it clear that by the enemy's center of gravity we do not mean a source of strength, but rather a critical vulnerability.[40]

'Moral Force' and Will-Power

Such a vulnerability could take many forms, including that of morale. The morale of the fighting forces, which Clausewitz called 'moral force', had already been identified as a crucial variable by Machiavelli in his call for a citizen army as opposed to the *condottieri* or leaders of mercenary forces found in Renaissance Italy. In his 1804 draft paper on strategy, Clausewitz wrote:

> Machiavelli, who has a very healthy judgement on matters of war, claims that it is more difficult with fresh troops to beat a recently victorious army, than it is before [victory]. He gives several examples as evidence and claims quite rightly that the moral advantage outweighs losses.
>
> A number of things come to mind in this context. Victories take on very different guises, and in the case of a not quite decided war, moral advantages may be slight. I would go further even and say that an army that has not quite been defeated may have the moral advantage on its side if it seizes the offensive on the following day or even the night. If, however, a defeated army has really been beaten so that it retreats, no matter how much the enemy is weakened by its victory, no matter how strong the new forces on the defeated side, all would be in vain, as the pursuing army would turn every engagement into a new victory.[41]

Machiavelli had not been the only strategist in modern times to comment on morale. In the age of Enlightenment the Welshman Henry Humphrey Evans Lloyd (c.1718–83) had shown some interest in it.[42] Similarly, Berenhorst in his *Reflections on the Art of War* stressed morale of troops as an important factor. Napoleon himself was known for his dictum that 'in war, three-thirds [of what matters] are moral forces'.[43]

The Clausewitzian paradigm of war – resulting from the interplay between the trinity of government, army and people – in fact hinges on the question of morale. As the Clausewitz expert Raymond Aron wrote, 'Clausewitz's emphasis on moral forces results from his interpretation of war as a social activity in which men are involved as a whole people – army, military leaders, head of state – all interdependent, the moral union of the people and the sovereign constituting the ultimate foundation of the state.'[44] Clausewitz explained that an army would be considerably strengthened by drawing its courage, enthusiasm and commitment to a cause from the spirit of the people. If war was the business of the people, they would feel strongly about the cause, would be prepared to make sacrifices and would in turn inspire the fighting forces to make greater efforts. Clausewitz devoted several chapters of his Books III and IV of *On War* to the importance of 'moral categories'. The main factors of morale he listed as 'the talent of the commander-in-chief, the warlike virtue of the army and its popular spirit' (in the sense of identifying with the people and drawing its strength from the passions of the people).[45] In his lessons for the Crown Prince, Clausewitz wrote (my italics):

> The first and foremost important rule to observe . . . is to use our entire force with the utmost energy. Any moderation shown would leave us short of our aim. Even if we are likely to be successful, we should be unwise not to make the utmost effort in order to win . . . *The moral impression created by these actions is of infinite importance* . . . [46]

In *On War*, he repeated the importance of morale-boosters, such as even minor victories, and emphasised that good morale could compensate for physical exhaustion. In turn, defeat could undermine the strength of armed forces, and any distrust of the competence of the commander-in-chief could erode the fighting spirit of an army more fatally than physical factors.[47]

Clausewitz was aware of the importance of morale on both sides of the war – among one's own forces and population, and among the enemy. As he wrote (again, my italics), 'Victory does not consist only in the conquest of the battlefield, but in the destruction of the physical *and moral* fighting forces.'[48] Ludendorff was thus quite wrong when he alleged that in *On War* Clausewitz 'has made no reference whatever' to 'the spiritual and psychical forces of the nation'.[49]

Again, we find that Clausewitz's ideas travelled. In his *Transformation of War* of 1911, Colonel Colin emphasised that the most important factor in achieving victory was 'patriotic passion'; on the other hand a crucial factor in dooming an army to defeat was 'the decline of national sentiment'.[50] 'Offensive spirit' and 'moral force' are Clausewitzian concepts that his French readers found particularly fascinating.[51] Moral force in turn was associated with the offensive. Colonel Hubert Camon warned his readers not to follow Clausewitz's teaching concerning the superiority of defence. Nevertheless, for Camon, Clausewitz's great worth lay in his emphasis on the moral element of war.[52] Another French admirer, Captain Georges Gilbert (1851–1901), had also been introduced to Clausewitz in his courses at the *École supérieure de guerre*. He appreciated Clausewitz's emphasis on moral force and boldness, and, exceptionally, his belief that in the last instance the defensive was stronger than the offensive, adding that a mind of such depth could hardly be mistaken. Otherwise he exhorted his readers to pay attention to Clausewitz's inspiration, Napoleon, famously pronouncing that the Germans 'have patiently developed the Napoleonic traditions which had been forgotten' in France.[53]

In 1902, Gilbert interpreted the Boer War as proving that morale was most important: a superior force had not been able to overcome the Boers easily because of the latter's superior morale.[54] Frédéric Culmann, another strategic commentator, agreed with this interpretation and applied it to the Russo-Japanese War, where he thought Japanese morale decisive and the power of defensive fire of secondary importance.[55] General Négrier commented after that war: 'Now it is everywhere recognised that with modern armaments the individual worth of the combatant has never been more important.' In view of the French inferiority in numbers compared with the Germans, 'This must comfort our hearts . . . Numbers no longer decide victory . . . A certain numerical inferiority does not trouble our soldiers.'[56]

And although tactics, as a contemporary Frenchman, Colonel Langlois, observed, changed every ten years, the constant was to rely on France's historic *élan*, on its moral force – a point for which ample confirmation was found in Clausewitz's writing.[57] Douglas Porch has commented: 'Clausewitzian "moral force" seemed an excellent cure for all of the ills which afflicted the French army – political divisions, deficiencies in armaments, lack of doctrine'[58] and, very importantly, when compared with Germany, inferior numbers. In the words of Foch,

> We, the French, possess a fighter, a soldier, undeniably superior to the one beyond the Vosges in his racial qualities, activity, intelligence, spirit, power of exaltation, devotion, patriotism . . .
>
> One hundred thousand men suffer ten thousand casualties and confess themselves beaten: they retreat before the victors who have lost as many men, if not more. Moreover, neither the one side nor the other knows, when retiring, either what numbers they have lost themselves or what the casualties have been on the opposite side (General Cardot). It is not, therefore the physical fact of having sustained losses, which makes them yield, and withdraw, give up fighting, abandon to the enemy the disputed ground the conquest of which marks the beginning of victory.[59]

Conservative French circles, which included the majority of military men, were sceptical of the achievements of the French Revolution. They too preferred an interpretation of the victories of the Revolutionary Wars which put the emphasis on French national character to one emphasising political and philosophical ideals. For example, in 1921, General Palat published a summary of *On War*. This included the following passage on war as a function of politics:

> The crushing of the best armies of Europe by those of the [first French Republic] and of the [Napoleonic] Empire owes less to the influence of politics on war than to the mistakes of those policies. It is not through new ideas, through procedures which were hitherto unknown, that the French changed the conduct of war, or where one has to look for the causes of so many astonishing actions, but in the [French] national character, the new [French] social state, the government and the organisation of France. The foreign cabinets thought they could dominate a situation the gravity of which they did not suspect. They put up weak defences [literally 'means'] in the face of forces which were going to crush them. Those were the political mistakes.[60]

The French were not alone in their emphasis on morale: in his pessimistic little booklet published in 1937, a retired German general, Horst von Metzsch, invoked Clausewitz's emphasis on the *Geist* (literally 'spirit', but here 'morale') which must not be found wanting when taking up arms. He quoted Clausewitz's confessional memorandum:

> Where the disparity of power is so great that no self-limitation imposed on our aims could protect us from going under, or where the probable duration of danger is so great that not even the most sparing use of forces can allow us to achieve our aim, there we have to aim the tension of our forces to execute a single desperate stroke. The side thus cornered will hardly expect any further relief . . . and will put his entire, ultimate faith in the moral superiority which desperation gives to the brave. He will regard the greatest daring as the greatest wisdom, may give his hand to ruses and, where no success is forthcoming, he will find the right of future revival in an honourable defeat.

He added: 'This thought of Clausewitz's is the central idea of the present work.'[61]

Other German military writers were also impressed by Clausewitz's ideas on the importance of psychological factors and morale. As Ulrich Marwedel has found, whenever morale or moral factors were cited in German strategic thinking or military comments around the turn of the nineteenth century, Clausewitz was invoked.[62] By the late twentieth century, much of Clausewitz's teaching on the importance of morale had become a commonplace and found its recognition in the military teaching of most Western countries. Morale for him was connected to will-power, another concept not invented, but popularised by him.

In Clausewitz's mind, there was a close link between morale and will power. Already Joseph Comte de Maistre (1754–1821) had written, 'A battle lost . . . is a battle that one *thinks* one has lost.'[63] Clausewitz the Realist described war as a struggle between two wrestlers:

> Each of them seeks to force the other to fulfil his will through physical force; his immediate purpose is to throw the enemy to the ground and thus make him unable to resist further. *War is thus an act of force to compel the enemy do our will* . . . Force, that is physical force . . . is thus the means, and the purpose is to impose our will upon the enemy.[64]

Clausewitz therefore at the end of his life formulated this more general definition of war. For, once Clausewitz had acknowledged the wide range of possible manifestations of war, and the ways in which it could escalate, his idealistic definition of war as something aimed at annihilating the forces of the enemy was no longer adequate. According to this new definition then, victory is the achievement of certain well-defined war aims which the enemy had sought to deny, i.e., *victory is succeeding in compelling the enemy to do our will*. If the war aim is limited, as in the wars of the *ancien régime*, the achievement of the limited war aims constitutes a full and satisfactory victory. If the aims, like those of Napoleon, were unlimited, then only a comprehensive defeat of the enemy's forces was an acceptable outcome. Analysed in these terms, Clausewitz's ideal wars, the wars of Napoleon, differed so much from the pre-Napoleonic wars for two reasons: one, because Napoleon had unlimited war aims; and two, because he broke through the socio-economic limitations of the professional army by harnessing popular enthusiasm to the war effort.

We have already seen in our discussion of morale that Clausewitz further recognised that war, this act of force, affects minds as well as bodies – and here is the crucial recognition of that psychological dimension of war.[65] In the *Principles of War for the Crown Prince*, Clausewitz listed the conquest of public opinion among the 'three main purposes of the conduct of war'.[66] His emphasis on the psychological dimension of war was closely connected, of course, with his emphasis on the desirability of the involvement of the people, with the mobilisation of the entire population for the war effort, when 'great interests' are at stake. So battles not only affect the balance of forces in real terms: their outcome also leaves important impressions on the participating soldiers, on the population at large, and on any other observers. Even Clausewitz the Idealist in Book IV explained that he thought of the 'main battle' primarily in terms of its psychological effect on the enemy ('it is no mere mutual murder and its effect is more the killing of the enemy's courage than of the enemy's warriors'[67]).

Some earlier advocates of the arming of the people in the German-speaking lands saw the danger that such a war might spin out of government control and counselled against it. Others were undecided as to whether it would bring more advantages or more problems.[68] Moltke echoed Clausewitz's language on the contest of wills in battle:

> Victory in combat is the most important moment in war. It alone breaks the will of the enemy and forces him to submit himself to ours. It is not in general the conquest of a piece of land or of a strong point, but alone the destruction of the enemy's forces which will be decisive. This is therefore the chief objective of the operation.[69]

Clausewitz's use of language spread throughout Western military literature. A generation later, Colmar von der Goltz elevated 'will' to greatest prominence along with military genius.[70] At much the same time Clausewitz's greatest French admirer, Foch (drawing also on Cardot and Joseph de Maistre who had also used this Clausewitzian terminology) wrote:

> ... *Victory* means *will*.
> In order to reach [the battle's] *end* – which is the imposing of our will on the enemy – modern war uses but one *means*: the destruction of the organised forces of the enemy. That destruction is undertaken, *prepared*, by battle, which overthrows the enemy, disorganises his command, his discipline, his tactical connections, and his troops *as a force*. It is carried out by the pursuit, in the course of which the victor utilises the moral superiority which victory provides over the vanquished, and tears to pieces, finishes off, troops already demoralised, disorganised, no longer manageable – that is, forces which are no longer a force. What we are considering now is the act of war, the means of overthrowing the enemy and of securing victory. Therefore: War = the domain of moral force. Victory = moral superiority in the victors; moral depression in the vanquished. Battle = a struggle between two wills.[71]

And he concluded further:

> If the will to conquer is necessary to offering battle with any chance of success, it is criminal in the commander-in-chief to deliver or accept battle without possessing that superior will which must provide direction and impulsion for all. And if battle is thrust upon him by

circumstances, he must decide to give battle, to fight, in order to conquer in spite of it all. On the other hand, one must not fight for the sake of fighting.[72]

Again, we find that in the twentieth century this Clausewitzian terminology and the concepts it conveyed became increasingly common-place, and were often used without a direct reference to Clausewitz. Changing the enemy nation's will through crude force became a key aim of several early fans of air power, including General Giulio Douhet in Italy and Sir Hugh Trenchard, 'father of the RAF', in Britain. It became one of the guiding principles of the practice of air warfare in the Second World War.[73] The Service Regulations Manual of the German Air Force of 1936 deduced the following missions: 'It is the task of the armed forces in war to break the enemy's will. The will of the nation finds its strongest incarnation in the armed forces. It is thus the primary task in war to fight down the inimical armed forces.'[74] And in the second half of the twentieth century, the French strategic thinker General André Beaufre deduced his own definition of strategy as: 'The art of the dialectic of wills using force to resolve their conflict' in order to 'convince' the adversary 'that to engage or pursue the battle is pointless'.[75] By the end of the twentieth century, the heritage of Clausewitz's contest of wills was so widespread as to be taken for granted.

Economy of Forces

On another point, Clausewitz's teaching at best led to confusion. He devoted a brief chapter of Book III to the 'economy of forces'. While he usually hesitated to give a rule of thumb, here he pronounced that the commander-in-chief should 'always ensure that all his forces join in' with his operations, that none should stay idle:

> If he has armed forces in a place where the enemy does not engage them sufficiently, or if he lets a part of his forces be on the march, i.e., 'be dead', while fighting with the enemy, that is poor husbandry. In this sense there is such a thing as a waste of forces, which is worse than its inappropriate application. Once action is the order of the day, it is a prime necessity that all parts act, because even the most inappropriate action keeps a part of the enemy's armed forces preoccupied and tied down, while inactive forces are simply neutralised for the time being.[76]

In other words, Clausewitz here did not advocate that one should economise with the forces one sends on a military operation, but, to the contrary, emphasised that one should bear down on one's enemy with all the forces one had at one's disposal. He contradicted himself in the following book, however, when he wrote:

> The smaller the part of the armed forces who have actually fought, and the larger the part which has contributed to the decision through its mere presence as reserves, the less the adversary can wrest victory from our hands again by fielding a new force. The commander-in-chief and the army which have been most successful in conducting the engagement with the highest economy of forces and which have brought to bear everywhere the moral effect of strong reserves take the surest way to victory. In recent times one has to concede to the French, particularly when led by Bonaparte, great mastery in this area.[77]

Clausewitz's views on the economy of forces are thus open to a variety of interpretations, and, unsurprisingly, attracted the French. Colin reflected on Clausewitz's advocacy of the economy of forces: 'People often neglect what Clausewitz calls the economy of forces "in time" – that is to say, they decide to bring all the forces at their disposal to bear in the same theatre of war, but *successively*.'[78] Foch commented,

> The principle of the economy of forces . . . is the art of pouring *all* one's resources at a certain moment on one spot; of making use there of *all* troops, and to make such a thing possible, of making those troops permanently communicate with each other, instead of dividing them and attaching to each fraction some fixed and invariable function; its second part, a result having been attained, is the art of again so disposing the troops as to converge on, and act against, a new single object.
>
> Again: the Economy of Forces is the art of making the weight of *all* one's forces *successively* bear on the resistance which one may meet, and therefore of organising those forces by means of a system.
>
> The *necessity* of this principle was felt from the outset of the wars of the Revolution – because national wars deal with large numbers.[79]

For Foch, 'the starting point of modern war is the *working of the masses*, tending to some *common* action in which the largest possible number of forces are to take part. Such common action was submitted by Clausewitz to a twofold condition: *union of forces in time and in space*.'[80] Foch also took on board the importance of surprise which Clausewitz had treated in Book III along with the strength in numbers.[81]

Even if Jehuda Wallach's criticism rings true that Foch's understanding of Clausewitz was by and large superficial,[82] it is through Foch that Clausewitz's more practical writing has become popular.

The Americans, too, caught the bug of wanting to reduce military thinking to a set of memorable 'principles of war' (an echo of Foch?). The *War Department Training Regulations No.10–5* of 1921 (the ancestor of the Field Manual 10–5) listed them in the following way:

a. The Principle of the Objective.
b. The Principle of the Offensive.
c. The Principle of Mass.
d. The Principle of the Economy of Force.
-e. The Principle of Movement.
f. The Principle of Surprise.
g. The Principle of Security.
h. The Principle of Simplicity.
i. The Principle of Co-operation.[83]

'Digested' by Foch, so to speak, several of these principles can trace their ancestry to the headings of sub-chapters of Book III of Clausewitz's *On War*. Among them, 'Surprise' and 'Economy of Forces' are directly Clausewitzian, in all their paradoxical tension with Clausewitz's emphasis on superiority in numbers (the larger the forces, the more difficult to strike with surprise).

Friction and Chance

There is much less scope for misinterpreting Clausewitz's other key features of battle. His concept of 'friction' is regarded as extraordinarily helpful in explaining the difference between war plans and reality.[84] The following Clausewitzian lines are often-quoted: 'Everything in war is very simple, but the simplest thing is very difficult. These difficulties pile up and create a friction which no one can quite imagine who has not experienced war.' Similarly, 'Friction is the only concept which quite generally fits the difference between real war and war on paper.'[85]

Action in war is a movement in a constraining medium. Just as it is hardly possible to perform the most natural and easy movement, such as simple walking, with any ease and precision in water, one cannot even achieve a middling standard of performance in war with ordinary forces. For this reason, the real theoretician seems like a swimming instructor, who teaches movements on dry ground which are necessary for [locomotion] within the water, movements which will seem grotesque and exaggerated to those who do not take the water into consideration. But for the same reason theoreticians, who have never dived into the water ... make impractical and pointless [suggestions], as they only teach what anybody is able to do anyway: walk.[86]

Clausewitz's answer to the problem of friction was military discipline, and the repeated practising of routine actions by the armed forces.[87]

Clausewitz's words were already taken up by F. Constantin von Lossau in his book *Ideals of Warfare* where he wrote 'that the basic idea of any operation in war is mainly quite simple, but very difficult in its execution.'[88] From then onwards, many writers took up the idea of the paradox of simplicity and difficulty of military operations which Clausewitz had elaborated so picturesquely in his Book I in the section concerning friction.

Related to friction is Clausewitz's concept often referred to as 'the fog of war', with which he explained the difficulties of real warfare: 'Finally the great uncertainty of all facts presents a peculiar difficulty in war, because all actions take place in something virtually akin to dusk, which in addition, like fog or moonlight, gives objects an exaggerated size and a grotesque view.'[89] Even during America's Afghanistan war at the end of 2001, this term was used to explain bureaucratic secrecy and inability to adopt a more open information policy.[90]

But the emphasis of luck, chance or accident in war, which we have already encountered in Clausewitz's second 'trinity', stands out. He compared war with a 'game of cards', and called it 'the realm of chance'.[91] He wrote that no war plan outlasts the first encounter with the enemy, a view that was echoed by Moltke.[92]

Such concepts were easy for military thinkers of different countries to digest, and often, the role of Clausewitz in spreading them has been forgotten. This is not the case with the big debate as to whether offensive or defensive action is the stronger, as we shall see in the following chapter.

5

The Defensive-Offensive Debate, the Annihilation Battle and Total War

Clausewitz has been accused of being the 'unremitting proponent of offensive strategies', and the 'apostle of total war'.[1] In this chapter we will examine the blame that is often laid at the feet of Clausewitz for the *offensive à outrance* practised virtually on all sides during the First World War, and his association with the war aim of the annihilation of enemy forces and with total war, pinpointing Clausewitz's true position in the respective debates.

DEFENCE AND OFFENSIVE

Clausewitz: Defence as the Stronger Form

Starting with the allegation that Clausewitz was exclusively a proponent of offensive strategy, we can simply note that the opposite is true: he proclaimed defence to be the stronger form of fighting. Clausewitz's own thinking on defence and the offensive is close to that of his old teacher, Scharnhorst. According to Scharnhorst, defence contained within itself the offensive, as the first rule of defence was 'not merely to defend oneself, but to attack'.[2] And he added:

> An army that is to defend a province achieves its purpose most surely if it annihilates the inimical armed forces tasked with the conquest of this province. Admittedly, circumstances may arise which force the defending army to avoid battle as long as possible . . . yet when the moment has finally come and it is forced to accept battle in some position or other, it is its main purpose to wear down the inimical armed forces as

much as possible. No way can it achieve this purpose, however, if it only counters the strokes of the adversary, without striking back. Therefore it must combine the attack with the defence or must even proceed offensively in the first place; in any case it can only expect salvation from the offensive. One can therefore rightly say: there is no tactical defence.[3]

Again, this idea was not Clausewitz's alone. Besides Clausewitz, Constantin von Lossau in his book *War* of 1815 largely shared Scharnhorst's views on the subject. He saw the offensive as central to war, 'because even when defending oneself, one can go onto the offensive'. All the efforts of the party that has been attacked have to be directed to countering the attack and then launching the counter-attack.[4]

But as early as 1804, Clausewitz wrote in his fragment on strategy:

What is a defensive war? One in which nothing is done? No, that would be a unilateral war. However weak one is compared with the enemy, the whole war cannot be of a defensive nature. It would be better to count on one's luck and desperation and to pretend that one's weak forces are strong enough to accomplish something than merely to counter the adversary's strokes. A general defence would need more armed forces than an offensive . . .

Our intention must never restrict itself to a pure defence. One always has to nurture the idea of attacking [the enemy], and we must only allow ourselves to be pushed into a defensive posture by circumstances. Just as a well-ridden horse must never lag behind his reins, the general must never do less than circumstances permit, and must always be led a little further by his own initiative.[5]

In his lessons for the Crown Prince, Clausewitz dealt with both defensive and offensive warfare:

Politically speaking, a defensive war is a war which one wages for one's independence; strategically it is the kind of campaign in which I limit myself to fighting the enemy in a theatre of war which I have prepared for this purpose. Whether the battles which I wage in this theatre of war are offensive or defensive, makes no difference . . .

Defensive warfare . . . does not consist of waiting idly for things to happen. We must wait only if it brings us visible and decisive advantages . . .

The strategic offensive pursues the aim of the war directly, aiming directly at the destruction of the enemy's forces, while strategic defence seeks to reach this purpose indirectly.[6]

91

In *On War*, Clausewitz argued that the defence is stronger than the offensive.[7] In Book VI he wrote again, almost word-for-word, what he had written in his notes on strategy of 1804:

> What is the concept of defence? The fending off of a strike. What, therefore, is its characteristic? The awaiting of this strike. Whenever this feature is present, it makes an action defensive, and in war this feature alone distinguishes the defence from the attack. Because an absolute defence totally contradicts the concept of war, because only one side would be waging it, defence in war can only be relative, and this term must only be used for the whole, and not be applied to all its parts. An . . . engagement is defensive . . . as is a battle . . . [if they are characterised by] waiting and fending off . . . But because one has to return the enemy's strike if one wants to wage war oneself, then this act of attack within the defensive war comes under the main heading of 'defence' . . . The defensive form of warfare is thus not a [static] shield, but a shield which consists of skilful strikes.[8]
>
> . . . as . . . the defence has a negative purpose, that of preserving, and the attack has a positive, that of conquering, . . . one . . . has to say: the defensive form of warfare is in itself stronger than the attacking . . . .
>
> If defence is the stronger form of war, yet has a negative object, it follows that it should be used only so long as one is compelled to do so out of weakness, and be abandoned as soon as one is strong enough to pursue a positive purpose. Once one has used defensive measures victoriously, a more favourable balance of strength is usually created; thus the natural course in war is to begin defensively and end by attacking. It would therefore contradict the very idea of war to regard defence as its final purpose . . . A war in which one uses one's victory only to fend off [the enemy], and does not want to strike back, would be as illogical as a battle in which absolute defence (passivity) is used in all measures.[9]

He saw a similar dialectic in the attack, as 'the attack itself cannot be without an element of defence, and that is a defence which is much weaker' than that of the defending party.[10] And he repeated, 'Defence is nothing but the stronger form of war, by the means of which one will achieve victory, in order to proceed to the attack, i.e., to the positive purpose of war, once preponderance had been achieved.'[11]

In Book VI, Chapter 8, Clausewitz described it as a particular advantage for the defensive posture that it entailed waiting for the other side to strike, that it was thus reactive, and consisted of long periods of inactivity interspersed with periods of (re)activity.[12] But again and again, Clausewitz emphasised that such a strategically defensive reaction could consist of a tactical offensive.[13]

Oddly, despite the emphasis he puts on defence, Clausewitz did not especially advocate the defence of fortified places. He still recognised the function of fortifications as depots, as shields for large and rich cities, as covers for military camps, as bulwarks for an otherwise undefended province, as centres for the organisation of a militia (literally, 'the arming of the people'), and for the defence of strategic passes through mountains and rivers. But on balance, and drawing on the experience of the Napoleonic Wars, he thought fortifications no longer played the role they had before the French Revolution, since which Europe had seen 'large standing armies with mighty trains of artillery, mowing down in a machine-like fashion the resistance of individual points'; for that reason cities no longer dared offer resistance to sieges, as their burghers knew they would fall sooner or later, and if later, reprisals would only be the harsher for having resisted.[14]

Defence and Offensive in Prussia and Germany

In Prussia, Clausewitz's teaching on the superiority of defence was quite unwelcome, while his teaching that defence must not be passive was eagerly accepted. In general, German military writers of the nineteenth and early twentieth centuries at best conceded to Clausewitz that certain advantages could be gained from a defensive position during a particular tactical phase of the war. Many authors, however, including Bernhardi and von der Goltz, rejected his teaching out of hand, calling it outdated, and criticising Clausewitz's logic and the premises on which he had based his argument. They praised the advantages of the offensive war, which could make the best use of surprise, poetically calling it 'the stronger form of fighting', 'the greater vital force', and lauding the 'preponderance of the initiative', 'the capacity of mobilising the spiritual and moral forces of the army', 'the inner sense of power', the 'greater possibility ... of keeping one's main fighting forces together for a single strike'. In short, virtually all the German military writers of the period delivered an impassioned 'plea for the offensive', 'that first-born daughter of Bellona' the goddess of war, the 'arcanum of victory'. German soldiers were taught in simple terms that 'waging war means attacking', and 'only the offensive is the true conduct of war'. Efforts were made to direct the entire education of the army in peacetime towards the spirit of the offensive in wartime. Offensive warfare was portrayed as the only possible form of war for Germany, given its encirclement by so many enemies. The mere conservation of what had already been conquered was derided as almost petty. The 'attack in itself, for its own sake' was presented as preferable to anything else, even in adverse circumstances, because it contained 'a great strength and a great [promise] of victory in the battles of the future'. As von der Goltz put it: 'Our contemporary German way of war aims at a decisive battle, blow for blow, which we perceive as inseparable from the offensive. Tacitly the offensive idea forms the basis of all theoretical deliberations and largely also of practical exercises.'[15]

The elder Moltke, however, was still remarkably close to Scharnhorst's and Clausewitz's teaching, writing in a piece, 'Defence, Offensive and Fighting with Bayonets', between 1861 and 65:

> The defence is capable of choosing its position in such a way that the enemy has to advance across an open plain. Thus [the defence] will almost always have the time to choose with precision the distances between certain sectors of terrain or individual objects in order then to optimise the use of fire instantly.
>
> The advantages of the offensive [on the other hand] as such are clear and lasting. The party that acts on the basis of his own decision determines for himself the law to which the waiting party has to subject its counter-measures. The attacker has a clear aim, he himself chooses the way in which he wants to reach it; the defender has to divine the intention of the adversary, and to weigh the means for his defence.
>
> On the one hand we thus have the ready decision, on the other uncertainty and waiting. And in the last instance, the defender, too, has to turn to the offensive, if he wants to force the final decision. It is another question, however, whether we should not first exploit the obvious material advantages of the stationary fire-battle before we go over to the offensive.

He explained that infantry would stand a better chance of defeating an oncoming enemy if the latter had to advance a long distance under constant fire before engaging with the other side's infantry and their bayonets: clearly, the first party's infantry had nothing to gain from rushing towards the aggressor. If the oncoming enemy turned back while still under fire, however, he ought to be pursued and would make an excellent target for a victorious counter-offensive. Nevertheless,

> The offensive is not in the least just tactical. A clever military leadership will succeed often in choosing a defensive stance which is strategically of such an offensive nature that the enemy is forced to attack us there; only when casualties, shock and exhaustion have weakened him will we begin the tactical offensive. Thus the strategic offensive can easily be matched with tactical defence.

But again he explained that a long period in which the enemy could be worn down through constant fire should precede the classic clash of arms with the bayonet which would complete one's own victory.[16] And in 1869 he wrote in his 'Instructions for Higher Officers',

> In dubious cases and unclear circumstances, which exist so often in war, it will generally be advisable to seize and retain the initiative, rather than letting the adversary choose the rules . . .
>
> The much-debated question whether a battle should be conducted offensively or defensively cannot be answered in general terms. The . . . clearest sign of victory is that at the end of the battle we occupy the space which the enemy occupied at its beginning, and which he would not yield without being more or less defeated. Any victorious battle must therefore end in offensive action, but the question remains an open one as to whether it should begin with such action. The advantages of the offensive are widely known. Through it we impose the rules of action upon the enemy, and he must adapt his measures to ours, and must choose means to resist us . . . .
>
> The offensive knows in advance what it wants; the defence remains surrounded by uncertainty, and can [at best] guess at the intentions of the enemy.[17]

And in 1874, after the Franco-Prussian War, Moltke wrote, 'It is my conviction that the improvement of fire weapons has won a great advantage for tactical defence over the tactical offensive. We were always offensive in the campaign of 1870 and captured the strongest positions of the enemy, but with what sacrifices! It seems to me more advantageous to move on to the offensive only after having repulsed several attacks by the enemy.'[18] It is thus not quite true that, as Azar Gat claims, 'Moltke consciously disregarded Clausewitz's insistence that defence was intrinsically stronger than attack.'[19]

The preference of the late nineteenth and early twentieth centuries, however, was clearly for the offensive. Count Schlieffen recorded his opinion in 1893:

> The armament of the army has changed, to be sure, but the fundamental laws of combat remain the same, and one of these laws is that one cannot defeat the enemy without attacking . . . One must attack when one has the means in hand to do so, and one must attack so as to annihilate [the enemy] to the greatest possible extent.[20]

Schlieffen was in good company when he rejected any defensive planning *a priori*. Lieutenant General von Caemmerer long-windedly

dismissed Clausewitz's argument for the superiority of defence,[21] concluding that this part of Clausewitz's argument was 'a failure'.[22] Equally,
Friedrich von Bernhardi wrote, 'I have reached the conviction that
particularly in modern mass warfare the offensive is by far the superior
form of the warlike action', and therefore he sought to disprove the views
of 'the greatest military authority', Clausewitz.

> I thus reach the conclusion that the offensive way promises a greater
> chance of success than defence, and for that reason commends itself
> above all to the weaker power, as long as the balance of forces gives any
> hope at all for a good outcome. From a certain imbalance of forces
> upwards, even the most ingenious offensive cannot outweigh the forces
> of the enemy, but in such a situation defence is even less likely to bring
> about a good outcome. Thus I believe to have invalidated once and for
> all Clausewitz's fundamental tenet that defence is the stronger way of
> waging war. Clausewitz seems to be contradicting himself anyway, when
> he argues that he cannot conceive of a defence which does not include
> counter-offensive strikes, as such passiveness would not come under the
> heading of war for him.[23]

Thus a manual on the army and the fleet published in Germany between
1909 and the beginning of the Great War stated: 'Clausewitz sees "the
actual, absolute aim of the warlike act in the shattering [*Niederwerfung*] of
the enemy" and speaks of a limited aim [only] when the conditions for
shattering the enemy do not pertain . . . The party that is forced to opt for
pure defence and for holding on to its possessions has to opt for a limited
aim . . . [But] this limitation can only be intended for a short period, while
one is trying to make up for the advance gained by the enemy in his
mobilisation and force deployment.'[24] And a German General Staff
memorandum of 1902 read, 'We are not out for conquest, but seek merely
to defend what is ours. We shall probably never be the aggressor, always
the attacked. The swift success we shall need, however, can be achieved
with certainty only when we take the offensive.'[25]

In the inter-war period, the *Truppenführung* regulations (directions for
the leadership of the armed forces) of General von Seeckt, however, put
more emphasis on defence again, both in the forms of outright defence and
delaying defence that prepares for a counter-offensive. Clausewitz's ideas
on the importance of defence were thus reinstated. Indeed, the *Truppenführung* defined '*Abwehr*' (fending off an enemy attack) in two ways:
'defence', in which the defender is determined to fight a decisive battle,
and '*hinhaltendes Gefecht*' (delaying engagement), or '*hinhaltender Wider-
stand*' (delaying resistance), in which the defender seeks to avoid the
decisive battle (at least in that area). But while it can be argued that under

Seeckt, the *Reichswehr* returned to Clausewitz's idea of a flexible defence,[26] it ultimately does not matter what the *Reichswehr*'s leadership's preferences were, for Hitler was firmly wedded to the offensive.[27]

France and the 'Offensive à outrance'

German military authors were not alone in favouring the offensive. The political culture of all Europe in the late nineteenth century lent itself to martial posturing. From boys' magazines to military parades, from music for marching to the social status of military officers, Europe's peoples were conditioned to think in martial terms, to see war as the potential remedy for industrial society's woes, and nationalism as the new religion for the uprooted urban proletariat. Military writers on all sides dismissed or reinterpreted Clausewitz's views on the strength of defence.

In France, for example, there was an equally keen interest in the defensive/offensive debate. (Significantly, the second translation of Clausewitz's *On War* into French by Lieutenant Colonel de Vatry was originally confined to Books III to VI, as Books I and II were described as 'too essentially philosophical', and Books VII and VIII 'too much of a sketch compared with the preceding' books.[28]) Shortly before the Franco-Prussian war, the equipment of the French Army with the *Chassepot* (the rapidly firing gun), which was known to be superior to its Prussian equivalent, the *Dreyse*, had led, in 1867, to the adoption by the French of instructions which practically defined defence as stronger than attack in the light of this technical innovation: henceforth, assuming that all sides would have such weapons sooner or later, firepower should make offensives more wasteful in human lives than ever before, as troops engaged in the offensive would be shot down in even larger numbers by the defensively used *Chassepot* and its equivalents. The French military historian Eugène Carrias commented: 'This decision could only have fatal consequences, as the defensive stance is incapable of producing decisive results.' And he blamed the French officer corps for being too obedient in applying these instructions.[29] Many of his countrymen agreed: strategically, the Franco-Prussian War was, to Frédéric Culmann, 'a resounding confirmation of the superiority of the offensive . . . and of the powerlessness of defence and above all of passive defence.'[30]

The reality seems more complex, however, as both the casualties of the Franco-Prussian War and of the First World War showed that indeed any infantry offensive against well-defended positions was henceforth blighted by the enormous firepower of defending artillery. The Prussian success in the war of 1870–71 was also a function of mobility, morale and the surprise effect of the rapid deployment of German forces by rail. The adoption of the machine gun in the First World War was yet another quantum leap in the development of firepower and hence of defence.

But the spirit of the age remained offensive. In Captain Gilbert's review of *On War* in *La nouvelle revue*, he emphasised the writing of Clausewitz the Idealist, the annihilation battle, the major 'Napoleonic' battle and the relentless pursuit of the enemy. In his *Essais de critique militaire*, Gilbert homed in on and condemned Clausewitz's dictum that defence is stronger than the offensive, but recognised that Clausewitz did not write about passive defence, but a defensive offensive.[31] To that extent, Gilbert approved grudgingly of *On War*, adding, however, that for France only the offensive was appropriate in view of her national character. And generally, Gilbert warned against paying more attention to Clausewitz than to the French original who had inspired him, instead urging his readers to study Napoleon's campaigns directly.[32]

In the last decades of the nineteenth century, Gilbert and Maillard as teachers at the *École militaire* taught that the 'offensive is the true mode of warfare, it alone leads to victory'.[33] L. Maillard, who rose to the rank of Lieutenant-Colonel, wrote in his *Elements of War* (1891): 'The destruction of the enemy is the aim; the offensive is the means.'[34] As their colleague Derrécagaix formulated it, 'The offensive is the only way to go for a general who wants to win.'[35] And Joffre wrote in his memoirs: 'Under the direction of Foch, Lanrezac and Bourderiat, the young intellectual élite at the *École de Guerre* now threw out the divisive old doctrine' of the primacy of defence based on the experience of the Franco–Prussian War. But 'the value of the offensive was [in turn] exaggerated by this group.' One spoke of the 'mystique of the offensive', and the 'cult of the offensive', which Marshal Joffre admitted took a 'somewhat irrational character' after 1905.[36]

Henri Bergson, with his popular vitalist philosophy (and without much debt to Clausewitz), further fanned French belief in the need for *élan* in battle. And Colonel Grandmaison championed the concept of the *offensive à outrance*, the idea that France did not know anything other than the offensive. The offensive was thus seen by contemporary military thinkers as the better alternative to the static defence of fortresses, as practised by the French in the war of 1870–71. (Clausewitz, who had seen little value in the defence of places, could at least be cited favourably in this context.[37])

Military leaders throughout the Western world typically saw no third way between aggressive offensive and passive defensiveness, and Book VI of *On War*, in which Clausewitz set out his understanding of the virtues of defence, was eclipsed by this thinking.[38] The American analyst Brodie commented (on Foch, but also on all the Great War strategists):

With Foch war became altogether an end in itself. In this respect he reflects only too faithfully the bent of his entire generation of soldiers. Lost completely is the cast of thought that makes Clausewitz both timeless and profound, his constant awareness that war is a political act, fought for a purpose outside itself. True, Clausewitz had cautioned against unduly tempering violence lest it render one weak against the ruthless, but he nevertheless insisted that the end must govern the means. The generation that was to fight World War I remembered only the injunction against tempering violence. 'Blood is the price of victory' was the one dictum of Clausewitz they quoted everlastingly.

Less arresting, but infinitely wider, was the admonition they forgot: 'No war is begun, or at least no war should be begun if people acted wisely, without first finding an answer to the question: what is to be attained by and in war?' The influence of this question, Clausewitz insisted, 'manifests itself down to the smallest details of action.' Foch gives little indication in his writing of having thought about the matter at all.[39]

Just like his contemporaries', Foch's thinking – if possible even more than Moltke's – was dominated by the quest for a decisive victory in a great battle (what Clausewitz would have called 'main battle'). The writing of Clausewitz the Idealist furnished him with the appropriate quotations when he sought to articulate this obsession. From Clausewitz's interpretation of the French Revolution he borrowed the analysis of the need for the population's support for the war effort, and its importance for morale in the armed forces. Foch wrote in his *Principles of War*:

> Let us first of all establish this principle that, if it is wholly to fulfil the twofold object of being the *rational end* of strategical operations and the *efficient means* of tactics, battle cannot be merely defensive.
>
> Under that shape, it may indeed succeed in holding up a marching enemy; it prevents him from reaching an immediate objective. Such results are, however, exclusively negative. Defensive battle never brings about the destruction of enemy forces; it never allows one to conquer the ground held by the enemy (which after all is the only external sign of victory), therefore it is unable to create victory.
>
> Hence the conclusion that the *offensive* form alone, be it resorted to at once or only after the *defensive*, can lead to results, and must therefore *always* be adopted – at least in the end.
>
> Any defensive battle must, then, end in an offensive action, in a thrust, in a successful counter-attack, otherwise there is no result.[40]

Some, said Foch, think the outcome of battle depends on details, personalities or manoeuvres, but 'victory is the consequence of efforts . . . which . . . aim at a common goal . . . namely, at a decision, a conclusion which alone can provide victory.'[41]

One might have expected 'that the offensive French military strategy was shaped by the offensive political aim of recovering Alsace-Lorraine and not by military-operational considerations. In fact, however, the desire to recapture the provinces lost to Germany in 1870' initially 'had no major effect on French strategy'.[42]

As the expert on French military history Jack Snyder has demonstrated, French war plans gradually turned offensive in the early 1900s.[43] The French War Plans 1–7, drawn up between 1875 and 1886, were thus still defensive. But with the Napoleonic revival in the 1880s, The French War Plans Nos. 8 and 9 became more offensive, culminating in the infamous Plan 17 of 1913, drawn up under Joffre and operational from 15 April 1914, which encapsulated the cult of the offensive. It emphasised all-out attack on Germany for the reconquest of Alsace and Lorraine. A similar move was reflected in their military manuals. The Infantry Field Regulations of 1884 extolled 'the principle of the decisive attack, head held high, with no attention to losses'. The Field Service Regulations of 1895 called for breaking the enemy's will and for offensive action, resulting in decisive victories through rapid concentration of forces.[44]

As we have seen, the French believed themselves to be a temperamental people, suited for the offensive more than for any orderly defence. The cult of the offensive and *élan* became more peculiarly French under the shadow of the increasing material and manpower dominance of Germany, which France tried to compensate with will-power.

There were a few French thinkers who had an inkling of what the First World War would be like, including Captain (later Major) Émile Mayer, and François de Négrier, but they were largely ignored. Both judged from the experience of the Boer War that 'front lines had become almost inviolable'.[45] The Socialist leader Jean Jaurès in *L'Armée nouvelle* (1910) pointed out that Clausewitz had regarded the defence as stronger than the offensive, and he strove to 'free French thinking from the grip of German militarism'.[46] But these were minority views, another one being that of Admiral Raoul Castex, whose writing dated mainly from the 1930s and 1940s.

Two of the few second-hand references to Clausewitz in Admiral Castex's treatise on *Strategic Theories* stress the defensive-offensive nature of Clausewitz's ideas on defence. Castex quotes Clausewitz as saying 'The different objectives that one can propose to attain through war are positive ones, and therefore, only the offensive can achieve them', and Castex's statement that defence 'must not be represented as a shield but rather as an arm as quick to riposte as to parry'[47] would have met with Clausewitz's approval.

The Offensive in Other Countries

The spread of the cult of the offensive was not limited in time or space, and it assumed a considerable momentum both within and outside a Clausewitzian tradition. As we have seen in Chapter 4, the American *War Department Training Regulations No. 10–5* of 1921 listed 'The Principle of the Offensive' as second among its nine 'principles of war',[48] and at much the same time, the Italian air-power theorist Giulio Douhet felt that the lesson of the First World War was that defence on the ground would henceforth be almost unbeatable, while in the air the offensive would always win. The formula for success in the future would thus be to rely on air power to defeat the enemy.[49] Friedrich Engels and Lenin did not like Clausewitz's ideas on defence either: they described defence as the death of the armed insurrection, and noted that there were few examples of a victorious outcome of a defensive war.[50] Even before the Revolution, Lenin declared that the offensive would determine it, adding: 'Once the insurrection is begun, it is necessary to act with the greatest determination, and by all means, without fail, to take the offensive. Defence is the death of an armed uprising.'[51] And Frunze, one of the great fathers of Soviet strategy, wrote,

> The tactics of the Red Army were and will be impregnated with activity in the spirit of bold and energetically conducted offensive operations. This flows from the class nature of the workers' and peasants' army and at the same time coincides with the requirements of military art.[52]

Exceptionally the Soviet strategist A. A. Svetchin, writing in the 1920s about the *Evolution of the Art of War*, thought that the lessons of the First World War confirmed Clausewitz's idea that defence was the stronger form of fighting, as 'Defence in strategy has the opportunity of using the boundaries and depth of the theatre, which forces the attacking side to waste forces in order to strengthen the spaces and to waste time crossing it, and any gain in time is another plus for the defence. The defending side reaps where it sows ... since an offensive is often stopped by false reconnaissance data, false fears, and inertness.' Svetchin thought it odd that the admirers of Clausewitz on all sides of the conflict had failed to recognise his views on defence in their conduct of the war.[53]

Classically, however, the 1936 Soviet Field Regulation stated:

> Every war, offensive and defensive, has the aim of defeating the enemy.

But only a decisive offensive in the main direction, concluding with persistent pursuit, leads to a complete annihilation of the forces and means of the enemy.[54]

And other military regulations dating from the beginning of the Second World War repeated consistently that 'Offensive combat is the basic aspect of actions by the Red Army.'

Mao Zedong echoed Clausewitz's emphasis on the strength of defence and on the annihilation of the enemy army. Like Engels and Lenin he thought, however, that the offensive had to be the stronger form of war than defence:

It has to be stressed that among the aims of war the annihilation of the enemy is the main task and self-preservation comes second; for only if the enemy has been annihilated in great numbers, can self-preservation be achieved effectively. For that reason the attack is the supreme task as the main means of annihilation of the enemy, and defence is the secondary task, as it is the auxiliary means for the annihilation of the enemy and . . . for self-preservation. Admittedly, in the practice of war defence is used most of the time as main means, while the attack is only used as main means in the remainder of the time; yet if one regards war as a whole, the attack is supreme.[55]

From another side, Clausewitz was criticised, oddly enough, for being too offensive in his defensive approach. Deeply scarred by the horrible experience of the Great War, Liddell Hart criticised Clausewitz for arguing that even defence had to turn into an offensive sooner or later. Writing in 1937, he sought to persuade his contemporaries that a purely defensive stance was wise for Britain:

Military action should be ruled by its head: the national object. We may be drawn into war to defend our interests and ensure, in face of an aggressor, the continuance of liberal civilisation, those larger ideas which we epitomise when we speak of 'England'. To attain that object need not imply on our part a war *à outrance*. For the aggressor, aiming at conquest, the complete overthrow of the opposing forces and the occupation of the opponent's territory may be necessary to his success. But not for ours. Our object is fulfilled if we can convince the enemy that he cannot conquer.[56]

Liddell Hart became tainted with appeasement. But, as Azar Gat rightly remarked, one must say in his defence that he sought ways, humanely and legitimately, to escape his age's obsession with absolute war, and to get back to *limited* warfare, one not fought with the sole aim of annihilating the enemy, on the battlefield or beyond.

ANNIHILATION OR DECISIVE BATTLE

Clausewitz the Idealist and his Contemporaries

Clausewitz is often thought, mistakenly, to have favoured the offensive by all means because of his emphasis, in his idealistic phase, on the annihilation of the enemy's forces in battle. The writing of Clausewitz the Idealist, which emphasised the importance of what Clausewitz called the 'main battle' (later referred to by others as the 'decisive battle' or 'annihilation battle'), was by far his most popular in the second half of the nineteenth and early twentieth centuries.

We have seen in Chapter 2 how much Clausewitz the Idealist emphasised the need for annihilation of the enemy in his 'Strategy' of 1804,[57] in the *Principles of War for the Crown Prince*,[58] and in Book IV of *On War*.[59]

In Book VII, Clausewitz the Realist cautioned that the offensive loses its drive the further it advances. He gave several reasons, including particularly overstretch (if one occupies part of the enemy's territory, one is usually forced to occupy all the territory between one's own and that part, to assure supplies) and, closely connected, the long lines of supply.[60] On the basis of the same reasoning, and in a chapter that greatly impressed Lenin, in Book VI he developed the virtues of a defensive withdrawal into the interior of a country, as the enemy, in advancing, would thus weaken. Clausewitz emphasised that the defending party should in such a case *avoid* the 'main battle' or postpone it till the very last possible moment, as such a battle would lead to substantial losses for the defending party, particularly if it led to its defeat. Otherwise, the defending party could choose where to delay the advancing invader with small engagements, making good use of the local topography which the defender was likely to know better than the aggressor. Clausewitz cited as an example Napoleon's advance into Russia in 1812: the battle of Borodino led to massive losses on both sides, while the avoidance of battle and the subsequent French seizure of Moscow allowed the Russian forces to hold out and finally drive back the weakened French forces.[61] Clausewitz thus developed an important qualification to the quest for the 'major battle' for the defending party, and, notwithstanding Liddell Hart's and several Frenchmen's criticism, thought that an indirect approach, which he described as characteristically consisting of an 'embracing' or 'outflanking' movement,[62] could be the key to victory in an offensive battle.

Indeed, even in his earlier idealistic phase, he had written in Book IV:

No state should make its fate, that is its entire existence, depend on one battle, even if it is the most decisive one. Even if it is defeated, it can raise new forces, and the natural weakening of . . . any attack can lead to a reversal of the situation, or it can receive aid from the outside.[63]

Yet elsewhere Clausewitz the Idealist stressed the all-important nature of battle. He was not alone in this belief. While an earlier generation had described battles almost like a special form of ballet, arguing that they might be avoided, Clausewitz's generation of Prussian reformers thought otherwise. Von Lossau, in his book *War*, put the same emphasis as Clausewitz on its importance – battle alone could bring the decision, and postponing the decision, he argued, was contrary to the nature of war. Like Clausewitz in his idealistic phase, Lossau saw war in Napoleonic terms, writing that the supreme commander 'in cold blood regards war as a struggle for existence, and peace as possible only through the shattering [*Niederwerfung*] of the enemy, to whom one has to be able to dictate the laws of peace with the sword.'[64] As the poet Novalis (Georg Philipp Friedrich Freiherr von Hardenberg, 1772–1801) wrote in reaction to the French Revolutionary Wars, 'What is the battle? . . . The purpose of the battle is to annihilate the inimical army. It can be achieved through its attrition or its dissolution as an army.'[65] The call for the annihilation of the enemy's forces was thus neither unique to Clausewitz in his own time nor to his age.

The German Reception of the Decisive or Annihilation Battle

In 1854, a year after reading *On War*, Friedrich Engels wrote:

In war there is only one right political line [to take]: with the greatest speed and energy to apply oneself to beating the enemy and to force him to submit to the demands of the victor. If the allied governments do so, I shall approve; if they tie the hands of their military commanders and gag them, I shall speak out against it . . .[66]

Moltke was also greatly inspired by Clausewitz's treatment of the battle of annihilation, and his emphasis on the importance of battle. Much more than Clausewitz himself, he elevated the need for annihilation of the enemy's forces to the chief aim of all military operations.[67] Not least under the influence of Moltke, the vision of war that Clausewitz the Realist had was largely ignored, and his idealistic writings were elevated to a widely held dogma in the Prussian and later German military. For Moltke, the aim in war was to achieve a quick decision:

> The character of the modern conduct of war is determined by the quest for large and quick decisions. The mobilisation of all the able-bodied, the size of the armies, the difficulties in feeding them, the expensiveness of the mobilised forces, the interruption of trade and commerce, industry and agriculture, and the armies' organised readiness to strike and the ease with which they are assembled, all this urges the quick termination of a war.[68]

In his youth, Moltke had been quite moderate in his thinking about war and peace. In 1841 he had written: 'We openly avow our support for the widely ridiculed idea of a general European peace', and in 1842, 'The further the social relations in Europe develop, the greater the dominating material interests are and the greater commerce and production grow, the more necessary undebatably is the peoples' need for peace.'

But gradually Moltke turned into a Social Darwinist.[69] In 1859, he yearned for war, 'for life and death', for 'a people's war', and in 1871 he even spoke approvingly of a 'war of extermination'.[70] On 11 December 1880, Moltke wrote his famous open letter to Professor Bluntschli in Heidelberg, in which he said, 'Eternal peace is a dream, and not even a beautiful one.'[71] In reply to a French member of the Society of the Friends of Peace, who reacted to this open letter by opining that every war was a crime, Moltke wrote on 10 February 1881:

> [War is] a last but perfectly justified means to defend the existence, the independence and the honour of a state. Hopefully this last resort will be applied less and less frequently due to the progress of culture, but no state can abstain from it entirely. For the life of man, indeed all of nature is a fight between that which is coming into being against that which is, and the life of entities of people is not much different. Who would question that every war, even the victorious one, is a misfortune for one's own people, for no conquest of territory, no milliards [of money] can replace human lives or can outweigh the grief of [bereaved] families. But who in this world can escape misfortune or necessity? Are not both, due to God's providence, conditions of our earthly existence? ... [Schiller's character] Max [in *Wallenstein*] says: 'War is terrible like heaven's plagues,/ yet it is good, is fate, like them.' And that war also has its beautiful side, that it brings forth virtues which would otherwise slumber or become extinct, can hardly be disputed.[72]

The cultural preference for doctrines of 'action' was one of the main characteristics of the decades leading up to the Great War. In the words of the politician Axel von Freytag-Loringhoven (1878–1942), 'in the soldier's life . . . action . . . must always be rated more highly than the thought'. Theory was derided as 'gymnastics of thought' (von der Goltz).[73] Wilhelm von Blume, who wrote *Strategie* (1882), Lieutenant-General Albrecht von Boguslawski, who wrote *War in its true Significance to the State and People* (1892), and Colmar von der Goltz in *The Nation in Arms* (German original 1883) all agreed with Moltke that war was inevitable and indeed desirable: in von der Goltz's words, 'Wars are the fate of mankind, the inevitable destiny of nations;'[74]

> Victory brings might, might riches, but prosperity luxury . . . The more civilised, the more wealthy a nation becomes, the greater the capacity for pleasure and indulgence. It shrinks from effort and comes gradually to value property and ease more highly than the brutal pursuit of war.[75]
>
> Germany . . . has become rich, and her riches increase daily. She grows in culture, but this growth of culture is unfavourable to the warlike development of her people . . . Present-day philosophy teaches free development of personality. Everything which stands in its way would be put aside . . . Involuntarily the question arises; will the spoilt multitudes . . . be willing to respond to the stern call to sacrifice life and property in defence of the Fatherland?[76]

Von der Goltz agreed with Moltke that 'war serves the end of politics best by a complete defeat of the enemy'.[77] Nevertheless, for von der Goltz, as for Moltke, the idea of annihilation was still very much concentrated on the enemy's armed forces. 'When we talk about annihilation', he wrote, 'what we mean is to put [the enemy] in a physical and moral condition that will make him feel instantly unable to continue the war'; the enemy's main army 'incarnates so to speak the adversary's opposition' and is thus the target for one's main military effort. Even then, he argued, one does not defeat the enemy by wiping him out, but by crushing his hope for victory.[78]

So, the cult of the annihilation battle reigned supreme in Germany. Clausewitz was time and again praised for being 'the first apologist for the idea of annihilation in war', for having been 'deeply imbued with the idea of annihilation' and for having 'elevated it to be the point of departure of the teaching on war'. Limited war aims and a limited form of war were described as anachronistic. Von der Goltz and Count Schlieffen agreed on this point with their French colleague J. Colin.[79]

The idea of annihilation was the only one stressed by the various writers of introductions to the new popular editions of Clausewitz's *On War*.[80] To quote but one, General Walter Reinhardt, Chief of the Army Staff in

1919–20, 'recognised in Clausewitz's teaching mainly his emphasis on annihilation'.[81] This reflects the fact that from the formulation of the Schlieffen Plan until the end of the First World War, military, not political, considerations dominated military planning.

One must admit that the Schlieffen Plan, foreseeing the attack on France through Belgium, had something in common with Clausewitz's own war plan against France of 1830. Here Clausewitz argued that a direct attack on Lorraine through Metz from Mayence would be very difficult, as there were a number of fortified towns that would either have to be besieged and taken or that could threaten the advancing German armies from their flank. Instead, Clausewitz recommended going through Belgium, which he also regarded as very valuable in its own right. He recommended a 'main battle' in Belgium to decide the campaign, and the subsequent seizure of Venlo, Maastricht (which he assumed would be seized by the Belgians at the beginning of hostilities), Liège, Brussels, Louvain, Ghent, Antwerp and finally Namur, lest the German forces' flanks be threatened from these. At the same time he thought that France, allied as it was with Belgium at the time, might attack the German lands by advancing through Belgium.[82]

As Jehuda Wallach speculated, 'It probably never entered [Schlieffen's] mind that he deviated in the slightest degree from Clausewitz's theory . . . Whereas Clausewitz recognised various degrees in the selection of war aims, Schlieffen confined his theory to one sole object: annihilation of the adversary.'[83] But, as the strategist Panajotis Kondylis has rightly remarked, Schlieffen's idea of war centred on the annihilation of the adversarial army on the battle field, not on a long war of attrition in which both sides' economies would be competing with each other, or in which the entire populations would be involved in the war effort – either by suffering from the economic blockade, or by working in the war economy. Schlieffen – like his counterparts in other countries initially – wanted a short, sharp war with a decisive victory, not a long exhausting campaign.[84]

The Germans were not alone in their obsessive quest for the decisive victory. Jack Snyder has shown that the ideology of the decisive battle, reached through offensive action, came to be shared by the Germans, French, British and Russians on the eve of World War I.[85] As Manfred Rauchensteiner cogently summed up, in the nineteenth century,

> the sole aim in war was to annihilate the mass of the hostile armed forces. But as gradually . . . all important states in Europe took note of the Clausewitzian teaching of the [importance of the] superiority in numbers, the armies of the European states grew steadily and thus returned to their point of departure in the early nineteenth century from

which they had turned away: the people's war. Although there were a few voices which preached against mass armies and saw the future in small, hard-hitting elite armies, Social Darwinism did not leave any leeway for such ideas. Social Darwinism made the formation of mass armies logical and moreover described the outbreak of war as inevitable. War was increasingly conceived of in absolute terms, and this absolute form was seen as inevitable.

As a result, the war aims of annihilating the enemy's armed forces 'were increasingly identified as the purpose of war.'

In the case of the First World War it was political leaders who made the decision to go to war, yet 'only a moment later it became clear that it was no longer politics which decided on the purpose of the war, but that the mechanism, indeed the automaticity of military planning, no longer allowed the formulation of a [separate political] purpose of the war.' Germany had to go to war with France not because political considerations dictated it, but because there were no alternative operational plans for its army. Austria-Hungary, which still followed a clear geo-political purpose in fighting Serbia, had no conceivable geo-political interest in going to war with imperial Russia. Thus both sides had no purpose other than to render the armies of the other side defenceless, a war aim that triumphed over any rational political purpose. No political purpose could justify the millions of deaths which the strategy adopted on all sides entailed; indeed, prior to its outbreak, the states that were determined to go to war had had little or no idea what they wanted to achieve by it.[86] In his work on the Schlieffen Plan, the historian Gerhard Ritter commented: 'The outbreak of the war in 1914 is the most tragic example of a government's helpless dependence on the planning of strategists that history has ever seen.'[87] Concomitantly, the First World War showed that only a war with a clearly stated limited political purpose, and with a limited employment of military means, could lead to limited demands on the peace negotiations that would in turn prove acceptable to the defeated party.

Delbrück and the Strategy of Exhaustion

Only one German writer of the period, the warfare historian Hans Delbrück, continued to stress the existence of two forms of strategy, following Clausewitz's somewhat misguided and short-lived espousal of a dualistic paradigm of war. In Book I, Clausewitz described three ways of winning a war: first, directly, through the crushing of the enemy's armed forces in the main battle; second, indirectly, through diplomatic efforts leading to the collapse of the enemy's alliances and the erosion of support for the principal enemy; and thirdly through 'the tiring out of the enemy

. . . What we mean by the term "tiring out" in the context of the fighting is the gradual effect of the exhaustion of the physical forces and of the will [of the enemy] through drawing out the action in time.'[88] This clashed, however, with Clausewitz's idea of the essence of war: in Book VIII, Chapter 8, he differentiated between the realities of war and the theoretical concept: 'While in reality the exhaustion [*Erschöpfung*] or rather the tiring out [*Ermüdung*] of the stronger party has brought about peace . . . this cannot in theory [*philosophisch*] be seen as the general and ultimate aim of any form of defence . . .'[89]

In Hans Delbrück's own words,

> Founded on the teaching of Clausewitz and leaning on my reading of the writings of Frederick the Great and Napoleon, I have established the doctrine that there is a dual nature of war and thus also of strategy. One can be called the shattering strategy [*Niederwerfungsstrategie*]; its main (almost exclusive) means is the destruction of the inimical armed forces in battle. The other sort can be called the exhaustion strategy [*Ermattungsstrategie*]; apart from the battle, it has a number of other means, which can be subsumed under the title manoeuvre. As the commander-in-chief moves incessantly between these two poles, one can say . . . that the function of the exhaustion strategy is dual and the shattering strategy is unipolar.
>
> Even in the unipolar strategy we won't find that there is nothing but battles. But everything else, manoeuvres, occupation of provinces, investment of good positions, even fortresses, lines of communication – all that is secondary compared with the great tactical decision, the battle . . . In the dual strategy, manoeuvres have a value of their own, even outside their relation to the battle. The dual strategy also stands under the highest law of the decision by arms, but mostly this decision is not actually brought about by fighting to the finish, but it remains suspended over the heads of the two manoeuvring parties. The dual strategy has its rightful place where the strength (or even the will) of the attacker is not great enough to adopt the unipolar shattering strategy.[90]

All wars of history could, Delbrück believed, be divided into either the category of 'shattering [*Niederwerfungs*] strategy' or into

> the exhaustion or wearing-down [*Ermattungs oder Zermürbungs*] strategy – obviously not in the sense that all wars of the world, under whichever circumstances they have been fought, could be fitted into these categories as into a scheme, but in the sense that these are the two basic forms, on which there have been endless variations and approximations.

The main forms, however, continue to exist. Alexander, Caesar, Napoleon, Gneisenau, Moltke belong to the former school [of shattering strategy]; Pericles, Hannibal, Gustavus Adolphus, Prince Eugene, Marlborough, Frederic the Great and Wellington belong to the latter [exhaustion strategy].[91]

While Delbrück had recognised that in the larger part of *On War*, Clausewitz discussed only the former, and that he had not completed his work. Delbrück noted that due to this fact, and due to the practice of warfare in the second half of the nineteenth century, the opinion had become established that the former form of strategy was true to the spirit of war, and that all historical deviations from it were 'more or less excusable or explicable imperfections'.[92] He tried to rehabilitate this second, more 'limited', form of strategy as a viable option in its own right, and interestingly added new elements to it. Yet even Delbrück, while arguing that the annihilation battle strategy and the exhaustion strategy were equally valid forms of conducting war, conceded that the shattering strategy was 'today the recognised and for present circumstances the only natural and permissible' strategy.[93]

While Delbrück had begun to study Clausewitz in the late 1800s, he wrote his *History of the Art of War within the Framework of Political History* between 1900–20. Although he was as bellicose as anybody when the First World War began, he was one of the chief critics of Ludendorff after the end of that war. Delbrück was criticised by contemporary German thinkers who argued that Clausewitz had written about limited war with limited political aims, but not limited strategy within such a war, continuing to think of the utter destruction of the adversary's fighting forces as the only sensible aim in a battle. In the 1890s, Delbrück's views were hotly contested by the predominant military faction (led by Friedrich von Bernhardi), and the military faction seemed to prevail, being more in harmony with the spirit of their age. As one member of that faction wrote, 'If Clausewitz had taught what Delbrück found in his writing, we would have to contradict [Clausewitz] on the basis of sound military practice.' He described 'the teaching of a dual strategy . . . as an unnecessary theoretical complication' which 'from the military point of view even presented a threat that had to be fought against.'[94] Oddly it was Delbrück in his discussion of attritional warfare, not the chief proponents of the annihilation battle, who conceived of the need for supplementary actions outside the realm of pure fighting, such as 'inflicting every form of economic harm through destruction, forced payments, disturbances of trade, in the case of sea powers mainly through blockades'; indeed he thought that by applying such 'manoeuvres' one might occasionally win a war without giving battle.[95] Such measures were more likely to affect the civilian population than an annihilation strategy focusing entirely on the battlefield. Looking at Falkenhayn's strategy at Verdun, Delbrück wrote:

> The attack on Verdun was no attempt to make a break-through; it was no battle, it did not aim to bring about a great tactical decision. If Verdun had finally fallen, it would obviously have been of great, especially moral, importance, but according to Falkenhayn, this success was not absolutely necessary. The purpose of this operation 'along the Meuse', as Falkenhayn wanted to call it, was the exploitation of the advantage of our encircling position in order to inflict many more casualties on the enemy in the continuous fighting, than we would suffer ourselves. The French were not to be defeated, but to be 'bled white'. Notwithstanding the pain, they had to defend Verdun, as its surrender would have meant an unbearable loss of prestige. But they had to suffer much more pain than us. This 'bleeding white' can be called an exhaustion- or wearing down strategy, which does not simply imply passive waiting, or bloodless manoeuvre.[96]

The First World War on the Western Front turned out more closely like the exhaustion strategy as Delbrück had conceived it than the annihilation strategy which all sides had hoped to apply when the war started. Nevertheless, Delbrück's ideas did not come to dominate military thinking in the interwar period, where the quest for a decisive battle and a decisive victory continued to haunt the minds of strategists on all sides. In 1942, the central office of bookstores catering for German soldiers on the front distributed a special book entitled *Annihilation or Exhaustion-Strategy?*,[97] which the author, Gert Buchheit, had only begun to write in 1939. He took the two concepts from Delbrück and argued that both had their virtues, depending on circumstances. But, on balance, the author tended towards a strategy of decisive battles, referring for support to Clausewitz's argument that a war loses its momentum the more it moves away from absolute violence, and dismissing Delbrück as not having taken Clausewitz's idea of an active defence[98] seriously enough. The annihilation battle is described as the shortest way to victory, and all round-about ways of exhausting the enemy do not obviate the need of a decisive battle to clarify the situation in the end. This book was truly representative of the thinking of contemporary writers and more important still of the planning of the *Wehrmacht*, which did indeed aim at quick, decisive victories over the enemy's armed forces until 1941–2, hoping to avoid a war of attrition, let alone the complete mobilisation of the economy. Even the extermination of the Jews, the most classic case of the annihilation of a 'nation', was only initiated in November 1941, and was not part of the initial German war effort.[99] And yet the concept of the annihilation of the entire enemy nation, not just of its armed forces, could be found in print much earlier, written by the prominent German military figure General Erich Ludendorff. In 1935 he wrote, 'For a morally strong people, the war decision lies solely in the victory on the battlefield and in the annihilation of the enemy Army and of the enemy nation.'[100] We will return to Ludendorff's thinking in our discussion of 'Total War' below.

Clausewitz in the English-Speaking World

According to Russell Weigley, 'The first English-speaking generation to read Clausewitz got the impression that he believed victory is the only object of war, that when the guns begin to fire, diplomacy abdicates to military strategy.'[101] Henry Spenser Wilkinson, in his publications during the Great War, described Clausewitz's account of war as 'the most trustworthy and the most adequate which we possess', forming 'the starting point of every fresh inquiry'. Wilkinson was enamoured with the battle of annihilation and disdained limited war: 'The war that aims at striking down the enemy by the destruction of his forces is that of the successful State; the war that tries to limit its aims, and therefore its exertions, is that of the defeated.'[102] Earlier, in 1907, N. F. Maude wrote admiringly of Clausewitz that he was the first to describe war as the most forceful form of human competition, which Maude linked to the Social Darwinist teachings which were to contribute so greatly to the outbreak of the Great War. 'What Darwin accomplished for Biology generally, Clausewitz did for the life-history of Nations nearly half a century before him . . . both have proved the existence of the same law in each case, viz., "the survival of the fittest".'[103] Like Maude, J. F. C. Fuller, writing after the Second World War, drew out the parallels between Clausewitz's emphasis on the need of a war-like spirit as a national characteristic and Darwin's teaching.[104]

Elsewhere, Clausewitz was either credited with British virtues or declared out of date on account of British inventions. The 'introduction' to Clausewitz published by Major Stewart Murray in 1909 had a preface by Spenser Wilkinson which credited Clausewitz with good old British common sense. But '(1) The improved net-work of roads, (2) Railways, (3) Telegraphs, wire and wireless, (4) Improved arms, (5) Aviation' and '(6) Universal service armies' made some of Clausewitz's teaching outdated.[105] Furthermore, Admiral Sir Gerald Dickens argued that strategic bombing invalidated Clausewitz's tenet 'that the subjugation of an enemy is best accomplished by defeating its armed forces in battle.'[106]

The Annihilation Strategy in France

The critical French student of Clausewitz's work, Captain Gilbert, mainly praised him for his emphasis on battle and called him the 'apostle of campaigns with decisive solutions'. The offensive, he thought, corresponded best to France's 'aptitudes' and was thus to be adopted as a strategy in the coming war.[107] Writing a decade later, Gilbert's compatriot Derrécagaix praised Clausewitz's emphasis on the physical destruction of the forces of the enemy with the employment of all available means, without any concessions to constraints imposed by international law or 'philanthropical' considerations, a principle Derrécagaix thought France had to espouse without any reservations.[108] War or battles of attrition are conceived by Foch also, as 'parallel battles': 'In the parallel battle, tactics attempt, consciously or unconsciously, to break the other party's resistance by slowly and progressively using up enemy forces. To this end, fighting is kept up everywhere. It is fed everywhere.'[109] As we have seen, Foch was a great advocate of decisive battle: 'Decisive attack is the supreme argument used by modern battle, which itself is a struggle between nations fighting for their existence, for independence, or for some less noble interest; fighting, anyhow, with all their resources and passions.' The 'decisive attack' was for him 'the keystone of battle'.[110]

Raymond Aron observed that

> Foch, like Colmar von der Goltz, did not distinguish between absolute war and actual war. The concept of absolute war drew him towards total war, i.e., the total mobilisation of resources towards a radical solution. It is worth emphasising these last words, because Foch sees the mass confrontation not in the shape of a prolonged struggle but rather, with a view to 1806 and 1870, as a great battle [*Entscheidungsschlacht!*], one in which the fate of nations is in the balance.[111]

And sounding very much like any of his German colleagues, Colin wrote: 'The attack is the normal mode of action in war. No one should be allowed to command armies who is not disposed by nature to take the offensive.'[112]

It is worth contrasting the way in which World War I was fought by men such as these with Clausewitz's views on war not being 'an act of blind passion': he thought it

> . . . dominated by its political purpose, the value of this purpose must determine the size of the sacrifices to be made for it, both in magnitude and in duration. Once the expenditure of effort exceeds the value of the political purpose, the purpose must be renounced and peace must follow.[113]

The Ghost of Clausewitz

While in Germany, France and indeed Britain the spirit of the age on the eve of the First World War welcomed only the views of Clausewitz the Idealist, after the experience of its slaughter, Captain Basil Liddell Hart criticised Clausewitz for being the prophet of the annihilation battle because he had 'analysed, codified and deified the Napoleonic method' – 'the destruction of the enemy's main forces on the battlefield'. Liddell Hart claimed that it was 'short sighted, if natural' to delude oneself that 'the armed forces themselves were the real objective in war'.[114] In his Lees Knowles Lectures of 1932–3 delivered at Trinity College, Cambridge, he called Clausewitz 'the Mahdi of mass and mutual massacre'.[115] 'Clausewitz ridiculed the idea that "there is a skilful method of disarming and overcoming an enemy without great bloodshed, and that this is the proper tendency of the Art of War" . . . It did not occur to Clausewitz that such an idea might be dictated by enlightened self-interest, but a desire to draw profit from war, not merely a gladiatorial decision. Nor did he pause to reflect that this idea had inspired the past masters of the art of war, who had translated it into practice with profit to their cause.'[116] He stressed that Clausewitz 'developed, if he did not generate, the idea that the destruction of the enemy's armed forces was the only true object of strategy. He made this a dogma without meaning to . . .'[117] He thought it ironic that

> In their blind pursuit of 'the one means' – the destruction in battle of the enemy's armed forces, the Germans spurned the chance not only of seizing ill-defended Paris but of occupying the unguarded Channel Ports . . . Their folly would shortly be capped by that of the Allies who, blinded by the same theoretical maxim, would stubbornly hurl themselves in hopeless assaults on the entrenched enemy . . . [118]

Otherwise, the English-speaking world did not resist the temptation to make the annihilation of the enemy the one and only desirable goal in war. The US Army Field Service Regulations of 1923 echoed Clausewitzian language on the annihilation battle: 'The ultimate objective of all military operations is the destruction of the enemy's armed forces by battle. Decisive defeat in battle breaks the enemy's will to war and forces him to sue for peace.'[119] And the 1939 US Field Manual 100–5 read:

> The conduct of war is the art of employing the Armed Forces of a nation in combination with measures of economic and political constraint for the purpose of effecting a satisfactory peace . . . The ultimate objective of all military operation is the destruction of the enemy's armed forces in battle. Decisive defeat in battle breaks the

enemy's will to war and forces him to sue for peace which is the national aim . . . [120]

Similar wording is found in the 1962 US Field Manual 100–5:

> Every military operation must be directed toward a clearly defined, decisive and attainable objective. The ultimate military objective of war is the destruction of the enemy's armed forces and his will to fight. The objective of each operation must contribute to this ultimate objective. Each intermediate objective must be such that its attainment will most directly, quickly, and economically contribute to the purpose of the operation. The selection of an objective is based upon consideration of the means available, the enemy, and the area of operations. Every commander must understand and clearly define his objective and consider each contemplated action in light thereof. [121]

Only in the 1968 edition of the Field Manual did the objective of 'the destruction of the enemy's armed forces and his will to fight' change to the 'defeat of the enemy's armed forces'. [122]

We have already quoted two examples of the Soviet interpretation of the term 'annihilation of the enemy' as the objective of any war. Tukhachevsky wrote:

> A field army is completely unconcerned with covering the entire space of the borders between states. It is concentrated in the decisive direction, and considers that the best means of all to secure its border is by annihilation of the army of the enemy. From the beginning of the war . . . the manpower of the one side seek the manpower of the other side in order to destroy and annihilate it. [123]

The Soviet Field Order of 1939 stated: 'We shall conduct the war in an offensive way and we shall carry it onto the enemy's territory. The Red Army will fight until the annihilation and total destruction of the opponent.' [124] And as late as 1987, Marshal Sokolov, the Soviet Defence Minister, explained that

> The Military Doctrine of the member-states of the Warsaw Treaty has a strongly developed defensive character. We will never be the first to start a war . . . Nevertheless, it has to be taken into account at the same time that the aggressor can only be finally *crushed* through determined offensive actions. It is thus of particular importance to be ready at all times to subject the aggressor to an *annihilating defeat*. The defensive

actions must be prepared and conducted so that we do not lose or risk any territory. Active defence must thus start on the border between NATO and Warsaw Pact.[125]

By contrast, the other principal heir of Lenin, Mao Zedong, was very restrained in his interpretation of 'the annihilation of the enemy's forces' as a war aim. Echoing Clausewitz's later writing in *On War*, Mao wrote, 'To annihilate the enemy means to disarm him or to rob him "of his ability to resist", not, however, to annihilate him physically down to the last man.' And he continued: 'Thus the annihilation of the inimical armed forces is the basis of all bellicose actions.' 'The game of war – self-preservation and the annihilation – constitutes the purpose of war and serves as the basis of all bellicose actions, which are all – from the technical to the strategic – penetrated by this nature.'[126]

CLAUSEWITZ AND TOTAL WAR

We have seen that, notwithstanding his realist revisions of 1827–31, Clausewitz the Idealist exercised a more lasting and deep influence on strategic thinkers until well into the twentieth century. It is hard to disagree with Jan Philipp Reemtsma who concluded that Clausewitz's work had made a great contribution to the idea of annihilation, 'and it will be difficult to acquit the book entirely from a share of the responsibility for its effect'.[127] But does that make him also an advocate of total war? To tackle this question, we must, as elsewhere, begin with definitions.

The American historian Roger Chickering has shown that there is a dominating 'narrative' of modern military history, the plot of which turns on the progressive intensification of war from the (limited) 'cabinet wars' of the eighteenth century to the Napoleonic Wars, the American Civil War, the Franco–Prussian War, the First World War, and finally, the culmination of 'Total War', the Second World War with Auschwitz and Hiroshima.

The narrative is underlain by philosophical assumptions drawn from several German sources. Total war, the *telos* of the narrative, represents in a common reading the realisation of what Clausewitz called 'absolute war,' its liberation from the restraints that the Prussian philosopher himself identified in calculations of policy and the 'friction' of combat. In a variation on this theme, total war represents an ideal type of the sort that Max Weber envisaged. The phenomenon can never be fully realised; it poses instead the absolute toward which the development of warfare is tending . . .

Whatever the uncertainties over the ontology of total war, consensus reigns, at least implicitly, on the characteristics that define it. Total war is distinguished by its unprecedented intensity and extent. Theatres of operation span the globe; the scale of battle is practically limitless. Total war is fought heedless of the restraints of morality, custom, or international law, for the combatants are inspired by hatreds born of modern ideologies. Total war requires the mobilisation not only of armed forces but also of whole populations. The civilians who labour on the home front are accordingly no less essential to the war effort than are the soldiers, nor are they less vulnerable to attack. The war aims and political goals of the belligerents are unlimited in total war, which accordingly ends only in the destruction or collapse of one side.[128]

The problem is that the term 'total war' has been used in a variety of ways, but never actually appears in Clausewitz's own writing. In his youth, however, Clausewitz seems to have anticipated it. During the anti-Napoleonic wars of liberation, Clausewitz hyperbolically described it in his confessionary memorandum as 'a war of all against all. Not the king fights the king, not an army another, but a people fights another and the people includes king and army.'[129] This seems to predict Ludendorff's ideas of 'total war' as being war 'directed not only against the fighting forces, but indirectly also against the nations themselves', demanding 'the entire strength of a nation' to be mobilised, 'since such a war is directed against it.'[130]

In *On War*, Clausewitz was less extreme than in his confession memorandum. His normative idea was that of 'absolute', not 'total' war. Let us call to mind again how he defined 'absolute war' in Book VIII.

War had taken on its *absolute* nature . . . under Bonaparte . . .

War had achieved this state of perfection at Bonaparte's hands . . .

Since Bonaparte, war again became the concern of the people as a whole, first on one side, then on the other, and took an entirely different nature, or rather closely approached its true nature, its *absolute* perfection. There seemed no end to the resources mobilised; all limits disappeared . . . War, freed from all its conventional restraints, had broken loose in all its natural force . . . In most recent times, in which war attained its *absolute* force, we find most of what is generally a true and a necessary feature of war.[131]

As the German historian Hans-Ulrich Wehler summed it up, Clausewitz distilled the concept of 'absolute war' from his experiences of Napoleonic warfare; it was first and foremost an empirically observed form of war, as he described it in Book VIII. It is only thereafter that he turned 'absolute war' into an abstract ideal, still asserting that Napoleonic warfare had *come closest* to the ideal, which led him in Book I to treat 'absolute war' as an ideal that could not quite be realised in the real world where friction and political limitations influenced it. Thus in Book I Clausewitz uses the expression 'absolute war' in a heuristic-hermeneutic sense.[132] In real life, however, wrote Clausewitz the Realist, 'without any inconsistency wars can have all degrees of importance and energy, ranging from a war of extermination [*Vernichtungskrieg*] down to a simple armed observation'.[133]

Europe had had its share of wars of extermination. Of these, Clausewitz had studied campaigns of the Thirty Years War, and the scorched-earth policies of Louis XIV in the *grand siècle* of the seventeenth century. In the following two centuries, the convention gained ground in Europe that civilian populations should be spared in war as much as possible. But such restraint was not practised in colonial wars, which, particularly towards the end of the nineteenth century and in the first half of the twentieth century, tended to develop genocidal characteristics. The American treatment of the native Indian populations, the Belgian atrocities commited in their part of the Congo, and the German massacres of the Hereros in German South-West Africa all bore traits of genocide.[134]

At the end of the nineteenth century, this disregard for civilian lives gradually superseded the more gallant approach that had characterised warfare in Europe for almost two hundred years. On the eve of the First World War, the French strategist Jean Colin, for example, had no taste for Clausewitz's distinction between limited and absolute war, at least in as far as this concept's application to Colin's own times was concerned.

> Clausewitz still considered that a war of limited success was possible in his day, but nothing proves this opinion to be justified. At all events it no longer seems to be possible in the twentieth century for European wars ... Without speaking of the passions that would animate most of the belligerents, the material conditions of modern war no longer admit the avoidance of a radical decision by battle. The two armies, occupying the whole theatre of operations, march towards each other, and there is no issue but victory. It is impossible to avoid the encounter, impossible also to seek it in but a half-success. It seems as though the distinction made by Clausewitz in the last century between absolute offensive and offensive with a limited objective is no longer to be made, at any rate as far as European war is concerned.[135]

The desire to avoid such an absolute war, with its carnage and misery, was thus, in Colin's view, neither possible nor reasonable. Colin himself was killed in 1917, in one of the absolute battles of the First World War.

And Colin was not alone in thinking Clausewitz outdated. Apparently ignorant of Clausewitz's confessional memorandum, Ludendorff wrote in 1935:

> In his work, *On War* . . . von Clausewitz, the eminent master of the science of warfare, rightly asserted that warfare is always an act of violence by means of which one state tries to bring another state under its power. In his theories concerning the attainment of such a goal, Clausewitz only thinks of the annihilation of the hostile armies in battle. This opinion has become an immutable principle for the conduct of all wars, and to bear it in mind is the first task to be carried out in [total[136]] war. What Clausewitz says with regard to the idea of an annihilation to be effected on the battlefield will, therefore, always remain of extreme importance . . . As for the rest, von Clausewitz's book belongs to a past development in history which has now been entirely superseded, nay, a study of his work is even calculated to produce a confusing and bewildering effect. *Today the time has passed when one could speak of the 'different kinds of wars', as Clausewitz has done . . .*[137]

The concept of 'total war' in the sense of a total mobilisation of one's own nation for the war effort, including putting one's entire economy on a war footing, was first written about in France during the First World War by Alphonse Séché[138] and Georges Blanchou.[139] The *term* 'total war', however, was possibly first used by Léon Daudet in his homonymous book *La Guerre Totale*.[140] In this tradition, the American use of the term, as Hew Strachan has shown, has concentrated primarily on the resources mobilised on one's own side, and on the means employed in war, not on the war aims.[141]

The First World War was indeed a step towards total war both in terms of resources drawn upon and effect on the populations. All sides mobilised larger proportions of the population (including the female population) than any war since the French Revolutionary Wars, and the enemy's civilian population was targeted by Germany with the first air raids ever conducted against towns and with the unlimited submarine warfare that began in 1917. For Ludendorff, then, writing in the mid-1930s, total war 'far from being the concern of the military forces alone, directly touches the life and soul of every single member of the belligerent nations.' Total war 'is the result also of the introduction of universal conscription, on account of an increasing population, and of the use of new means of warfare, the effects of which have become more destructive.' Total war 'can be waged only when the existence of the entire nation is actually being threatened, and the latter is really determined to wage such a war.' Furthermore,

War being the highest test of a nation for the preservation of its existence, a [total] policy must . . . elaborate in peace-time plans for the necessary preparations required for the vital struggle of the nation in war, and fortify the foundations for such a vital struggle so strongly that they could not be moved in the heat of war, neither be broken or entirely destroyed through any measures taken by the enemy.[142]

For Ludendorff, politics in peacetime thus had to be made subservient to the preparations for a total war; politics was thus the projection of war into peacetime, and the function of war, not the other way around.

While this resonated with most National Socialists, opposition to them centred not least on this very issue. General Ludwig Beck, Hitler's first chief of the army General Staff, in a daring speech made to a circle of colleagues and friends after his dismissal in 1942, argued that those who branded Clausewitz as outdated were wrong, as Clausewitz had conceived of the notion of 'wars waged by both sides to the full extent of their national strength'. According to Ludendorff's definition, 'total war' as a new form of war (which Beck saw realised in the Second World War[143]) was simply an increase and extension into new areas of the 'act of force'. Beck quoted at length from *On War*, Book III, Chapter 17, where Clausewitz speculated on 'the nature of today's wars', arguing that the mobilisation of male civilians to fight in militias was an innovation which he doubted future wars would abandon. Beck further referred to Clausewitz's tenets that 'it is impossible to introduce into the philosophy of war the principle of moderation without becoming absurd' and 'war is an act of force, and there are no logical limits to its application. Each side imposes its law upon the other; this gives rise to a reciprocal interaction which must theoretically lead to the extreme.'[144] Beck continued: 'These quotations lead us to say that Clausewitz has defined as characteristic of war in general and essential to its concept the continuous extreme application of force, just what is now claimed to be the special characteristic of "total war".' But, he continued,

> Ludendorff . . . expects politics to serve war as the highest manifestation of the people's will to live, thus inverting the relationship between politics and war (for even in peacetimes, politics is to serve the teaching of total war). For Clausewitz, by contrast, all wars, including those in which the political element seems to disappear entirely, are always political actions. War, he concludes, is . . . in all circumstances no independent thing but to be thought of as a political instrument.

What Ludendorff had asked for, and what was being put into practice by Hitler and the *Wehrmacht*, must have the most drastic consequences for the life of a people. [There is n]o group, no area of human activities, no natural riches of the country, no existing or acquired property, that would not be seized and exploited [to prepare] for war. But in this way the preparations for war will become an insatiable moloch . . . In time this will lead to the over-exploitation of men and things, of spirit and soul for the purposes of the one goal; all other expressions of life will by and by wither away or come to a standstill.

Another crucial difference between Ludendorff and Beck (and Clause-witz, on whom Beck based his critique of Ludendorff) was that while Ludendorff postulated the primacy of the German people and an extreme Social Darwinism that saw no place for the peaceful coexistence of several vigorous peoples, Beck called upon the Germans 'not to ignore the right of other peoples to a life corresponding' to that of the Germans, and 'to prefer, where at all possible, the honourable and useful mediation to the use of force on principle.' For Ludendorff, peace could only be constructed on the total destruction of the enemy; Beck, by contrast, thought that a peace could only last if it was accepted on both sides, could, in Bismarck's words, last fifty years, and would not, like the peace of Versailles, serve as the ignition for the following war. Total war, however, as an end in itself, could not possibly bring forth an equitable peace, and, in Beck's words, necessarily had to give birth to the next total war. Beck thus pleaded for limitations imposed upon war, not its total unleashing for its own sake.[145]

Beck was arrested and executed in the aftermath of the 20 July 1944 assassination attempt against Hitler. This made him a point of moral reference for post-World War II German thinking. Thus the West German historian Gerhard Ritter, like Beck, criticised the attribution of the concept of 'total war' (which he defined in Ludendorffian terms) to Clausewitz:

The conceptual confusion . . . results, in my view, from the ambiguity of the concept of 'total war'. You may, in accordance with Clausewitz, take it to be 'absolute war' which does not stop until total destruction of the opponent's armed forces is achieved or, on the other hand, you may – against what Clausewitz had had in mind – perceive this concept as a war which becomes an end in itself . . . which requires all resources of a nation, and which . . . ultimately destroys more than ever can be reconstructed: the whole economic and social structure of a society, not only that of the opponent, but also that of the belligerent party who had embarked upon it. *Clausewitz, in contrast to that, talks only about a war between armies, not between societies.*[146]

But a confusion of terms persisted. Not only in Britain, but also in France, Clausewitz was linked with 'total war'.[147] It was thus not only Liddell Hart who thought that Clausewitz's 'absolute war' was a 'fight to a finish theory which, beginning with the argument that "war is only a continuation of state policy by other means", ended by making policy the slave of strategy.'[148]

> Clausewitz looked only to the end of war, not beyond war to the subsequent peace.[149]
> Because 1870 was taken to be the vindication of Clausewitz, his theory of absolute war became fixed on the mind of military Europe. Proclaimed by soldiers everywhere as an indisputable truth, it was submissively accepted by a generation of statesmen dangerously ignorant of war.[150]

Going further still, John Keegan wrote that Clausewitz promulgated 'the most pernicious philosophy of warmaking yet conceived'; 'I call Clausewitz pernicious because his political philosophy underlies that of the totalitarian state' (giving Hitler's references to Clausewitz as proof of this). Thus, he continues, Clausewitz polluted 'civilised thought about how wars could and should be fought.'[151]

The protean use of the term 'total war' is at the root of this confusion. Martin Shaw follows Michael Howard in attributing to Clausewitz the idea of 'total war' equalling 'absolute war' in the sense of an unlimited explosion of violence while recognising that the National Socialists in Germany took total war to unprecedented extremes with genocide.[152] But James John Turner is one example of those who argue that Clausewitz, 'in defining his concept of "absolute war," grasped well the nature of the [Max Weberian] ideal type of total war.'[153] As Bernard Brodie rightly commented,

> It is ironical that some of the very quotations which are often cited to prove that [Clausewitz] was the prophet of total or 'absolute' war are wrenched from a chapter (Ch.1, Bk.1) in which he specifically insists that 'war is never an isolated act' and that the military method must always defer to the political object.[154] Clausewitz, ambivalent [concerning victory] as in many other respects, can be and has often been quoted out of context to demonstrate his vehement rejection of restraint on war.[155]

Nevertheless, as the military historian Jay Luvaas remarked, in the light of subsequent developments, it was difficult if not outright impossible to see Clausewitz's teaching as wholly innocent, even if we judge him to have been free of any genocidal intentions.

Whether this new 'total war' was what Clausewitz had in mind when he referred to 'absolute war' is another matter, but because his theory seemed to specify the destruction of the enemy and allowed for the possibility that war could approach its 'pure concept with all its rigorous implications', Clausewitz after 1918 was often regarded in an entirely new light.[156]

Reemtsma's views of this, quoted above, are very close to this conclusion, which is thus common to scholars in different countries. Nevertheless, had history taken a different turn, had limited wars become the norm for the late nineteenth and the twentieth centuries, we would see Clausewitz in a different light. It is hardly fair to blame the entire development of subsequent military history on him.

6

Taking Clausewitz Further: Corbett and Maritime Warfare, Mao and *Guerrilla*

We have seen in the previous chapters that few strategic thinkers applied Clausewitz's teaching in all its complexity. Be it Moltke or Schlieffen, Foch or Bernhardi, most ignored the evolution and the resulting inconsistencies of *On War*, favouring mostly the writing of Clausewitz the Idealist as it fitted into the spirit of the time. The most creative use made of the writing of Clausewitz the Realist, however, we owe to the pens of two of his disciples, Sir Julian Corbett – who applied them to maritime strategy – and Mao Zedong, who applied them successfully to the special case of the civil war of his Communist forces against the Chinese nationalists. Both merit treatment in their own right.

Clausewitz Gone to Sea: Sir Julian Corbett

Sir Julian Corbett was born in 1854, two decades after Clausewitz's death, and died in 1922; he thus belonged to the generation of military thinkers who wrote in the period leading up to the First World War, but his thinking went very much against the fashion of the times. Corbett first came across Clausewitz's writing in the work of G. F. R. Henderson in about 1903. Corbett adopted Clausewitz's distinction between absolute and limited war, and applied his thinking about war as an instrument of politics. But he drew further conclusions, namely that different principles apply to naval warfare than apply to warfare on land. Sea-power, Corbett argued, depends on the ability to transport troops by sea to theatres of operation. To make this clear, Corbett spoke of 'maritime strategy', which dealt with the use of both the navy and armed forces deployed on land by the naval forces, through whom land could be conquered, held or defended. He emphasised the need for naval policy to be co-ordinated with other efforts, diplomatic and military. Naval battles, he argued, are neither always necessary nor always decisive. He wrote that the study of history showed that 'the command of the sea is only a means to an end. It never has been and never can be the end in itself.'[1]

Concentration of forces in his view was a 'shibboleth' which was always repeated but simply did not always apply to sound naval strategy.[2] He thus departed from Clausewitz in some decisive points and established a different perspective on the theory of naval strategy, although he took Clausewitz as the point of departure. And this was the key point on which the Board of the Admiralty took issue with him. When they finally approved the publication of his key text, *Some Principles of Maritime Strategy* (1911), they prefaced it, saying 'Their Lordships find that some of the principles advocated in this book, especially the tendency to minimise the importance of seeking battle and forcing it to a conclusion, are directly in conflict with their views.'[3]

In 1906, Corbett had written the 'Green Pamphlet' ('Strategic Terms and Definitions used in Lectures on Naval History') for the First Sea Lord, Admiral Sir John Fisher, for teaching purposes in the Naval War College. It emphasised that overall command of the sea was

> not essential to all oversea expeditions . . . under certain conditions . . . it may not be the primary function of the fleet to seek out the enemy's fleet and destroy it, because general command may be in dispute, while local command may be with us, and political or military considerations may demand of us an operation for which such local command is sufficient, and which cannot be delayed until we have obtained a complete decision . . .
>
> From the above it will appear that 'command of the sea' is too loose an expression for strategical discussion. For practical purposes should be substituted 'control of passage and communication'.

Thus a war minister might intelligently ask his chief of staff of the navy, not, 'Have we got command of the sea?' but can we 'secure the necessary lines of communications from obstruction by the enemy?' Usually it was said that 'the primary object of the fleet is to seek out the enemy's fleet and destroy it.' Instead, it should be that 'the primary object of the fleet is to secure communications, and if the enemy's fleet is in a position to render them unsafe it must be put out of action. The enemy's fleet usually is in this position but not always.' (Elsewhere he interpreted 'usually' as 'nine times out of ten'.[4])

Some Principles of Maritime Strategy was essentially an elaboration of the 'Green Pamphlet' with added historical examples, and is where Corbett's main study of Clausewitz can be found,[5] adding to his Clausewitzian treatment of the Seven Years War. In *Some Principles*, which was written to counter the prevailing arguments among the British General Staff in favour of a large-scale troop commitment to the European mainland, Corbett reiterated the importance of fitting naval strategy in with all other efforts, economic, diplomatic, and military on land. He stressed that the annihilation battle, the battle to overthrow the enemy utterly, was not necessarily crucial, indeed sometimes was a distraction from the more important aim of stopping the enemy from using the sea for his purposes. Command of the sea did not mean victory in decisive battles, just the ability to act where one chose to. Strategic defence, if one could not have command of the sea, for Britain was still enough for survival.

Although he had Fisher's support, Corbett was a very controversial writer because of his views on decisive battles. Admiral Custance and Spencer Wilkinson, whom we have already encountered and who belonged 'to the hearty or offensive-at-all-costs school',[6] disliked Corbett's writing strongly.

Corbett built his *Principles* on the historical studies he had done – on Sir Francis Drake, whom he admired unashamedly, on Drake's successors, and on *England in the Seven Years War*, in which he used Clausewitzian concepts. In *England* he wrote that

> there may be moments in the most complex war when the destruction of the enemy's main fleet and the securing of the command of a certain sea may be of an importance so great and pressing that naval action may rightly be left free to concern itself with nothing else, and every consideration of diplomatic and military operations must rest subservient to naval strategy. When such rare moments occur, they are invariably so dazzling in their dramatic intensity as to dull our vision of what they really mean and how they were brought about. The imagination comes naturally to concentrate itself upon such supreme catastrophes and to forget that war is not made up of them . . . The current conception of the functions of a fleet is dangerously narrowed, and our best minds cramp their strategical view by assuming unconsciously that the sole function of a fleet is to win battles at sea. That this is the supreme function of a fleet is certain . . . but on the other hand . . . convenient opportunities of winning a battle do not always occur when they are wanted. The great dramatic moments of naval history have to be worked for and the first preoccupation of the fleet will almost always be to bring them about by interference with the enemy's military and diplomatic arrangements.[7]

Corbett thought that by studying history one could find certain patterns ('the normal') or theories which allowed a certain amount of prediction.[8] From this he extrapolated rules for what he chose to call 'maritime strategy':

> By maritime strategy we mean the principles which govern a war in which the sea is a substantial factor. Naval strategy is but that part of it which determines the movements of the fleet when maritime strategy has determined what part the fleet must play in relation to the action of the land forces; for . . . it is almost impossible that a war can be decided by naval action alone.[9]

And here followed one of his most famous dicta:

> Since men live upon the land and not upon the sea, great issues between nations at war have always been decided – except in the rarest cases – either by what your army can do against your enemy's territory and national life, or else by the fear of what the fleet makes it possible for your army to do.[10]

The practical implications for government decision-making he saw as follows:

> [T]he [naval] Staff must ask of [the ministers] what is the policy which your diplomacy is pursuing, and where, and why, do you expect it to break down and force you to take up arms? The Staff has to carry on in fact when diplomacy has failed to achieve the object in view, and the method they will use will depend on the nature of that object . . . War is a continuation of policy, a form of political intercourse in which we fight battles instead of writing notes . . . [Clausewitz] saw that real war was in fact an international relation which differed from other international relations only in the method we adopted to achieve the object of our policy.[11]

We are, he continued,

> . . . dominated by the idea that since the Napoleonic era war has been essentially a different thing. Our teachers incline to insist that there is only one way of making war, and that is Napoleon's way. Ignoring the fact that he failed in the end, they brand as heresy the bare suggestion that there may be other ways, and not content with assuming that his system will fit all land wars, however much their natures and objects may differ, they would force naval warfare into the same uniform . . .[12]

To assume that one method of conducting war will suit all kinds of war is to fall a victim to abstract theory . . .[13]

Following Clausewitz, Corbett suggested the substitution of 'positive' and 'negative' respectively for 'offensive' and 'defensive'. Corbett was just about the only writer of his generation who built on Clausewitz's logic. But by reformulating it, Corbett clarified Clausewitz's ambiguous writing on this point.

If our object be positive our general plan must be offensive, and we should at least open with a true offensive movement; whereas if our object be negative our general plan will be preventive, and we may bide our time for our counter-attack. To this extent our action must always tend to the offensive. For counter-attack is the soul of defence. Defence is not a passive attitude, for that is the negation of war. Rightly conceived, it is an attitude of alert expectation. We wait for the moment when the enemy shall expose himself to a counter-stroke, the success of which will so far cripple him as to render us relatively strong enough to pass to the offensive ourselves.[14]

Corbett identified a difference between the 'German or Continental School of Strategy and the British or Maritime School'. Unusually, he understood Clausewitz's intellectual development, namely that only Book VIII of *On War* really spelt out Clausewitz's late thoughts about limited and absolute war.

It is clear that Clausewitz himself never apprehended the full significance of his brilliant theory. His outlook was still purely continental, and the limitations of continental warfare tend to veil the fuller meaning of the principle he had framed. Had he lived, there is little doubt he would have worked it out to its logical conclusion, but his death condemned his theory of limited war to remain in the inchoate condition in which he had left it

He then rightly noted the intellectual shortcomings of *On War*:

. . . All through his work Clausewitz had in his mind war between two contiguous or at least adjacent continental States . . . [I]n that type of war the principle of the limited object can rarely if ever assert itself in perfect precision. Clausewitz himself put it quite clearly. Assuming a case where 'the overthrow of the enemy' – that is, unlimited war – is beyond our strength, he points out that we need not therefore necessarily act on the defensive. Our action may still be positive and

128

offensive, but the object can be nothing more than 'the conquest of part of the enemy's country.' Such a conquest he knew might so far weaken your enemy or strengthen your own position as to enable you to secure a satisfactory peace . . .

Even then, Corbett recognised that Clausewitz had struggled with the problem:

> He was careful to point out that such a form of war was open to the gravest objections. Once you had occupied the territory you aimed at, your offensive action was, as a rule, arrested. A defensive attitude had to be assumed, and such an arrest of offensive action he had previously shown was inherently vicious . . . Added to this you might find that in your effort to occupy the territorial object you had so irretrievably separated your striking force from your home-defence force as to be in no position to meet your enemy if he was able to retort by acting on unlimited lines with a stroke at your heart.[15]

It is useful to pause here to reflect on the change of the political landscape between the times of Clausewitz, at the watershed between *anciens régimes* and the rise of nationalism, and Corbett, who lived at a time when the ideal of the nation-state had become dominant throughout the Western world and was spreading over the entire globe. At the time of Clausewitz, the German princes' territories were scattered all over central Europe, but by the time Corbett was writing, consolidated states inspired by nationalism had formed throughout Europe, with contiguous 'national' territory. Furthermore, many European states had by now acquired far-flung colonies.

> If we are aiming at a limited territorial object the proportion of defence required will tend to be much greater than if we are directing our attack on the main forces of the enemy. In unlimited war our attack will itself tend to defend everything elsewhere, by forcing the enemy to concentrate against our attack. Whether the limited form is justifiable or not therefore depends, as Clausewitz points out, on the geographical position of the object . . . When he conceived the idea, the only kind of limited object he had in his mind was . . . 'some conquests on the frontiers of the enemy's country', such as Silesia and Saxony for Frederick the Great, Belgium in his own war plan, and Alsace-Lorraine in that of Moltke.

Corbett put his finger on the shortcoming of Clausewitz's idea of a limited war aim in an age when nationalism made the cession of even the most limited chunk of territory to another state unacceptable:

> . . . Such objects are not truly limited, for two reasons. In the first place, such territory is usually an organic part of your enemy's country, or otherwise of so much importance to him that he will be willing to use unlimited effort to retain it. In the second place, there will be no strategical obstacle to his being able to use his whole force to that end. To satisfy the full conception of a limited object, one of two conditions is essential. Firstly, it must be not merely limited in area, but of really limited political importance; and secondly, it must be so situated as to be strategically isolated or to be capable of being reduced to practical isolation by strategical operations. Unless this condition exists, it is in the power of either belligerent, as Clausewitz himself saw, to pass to unlimited war if he so desires, and ignoring the territorial objective, to strike at the heart of his enemy and force him to desist.

> If, then, we only regard war between contiguous continental States, in which the object is the conquest of territory on either of their frontiers, we get no real generic difference between limited and unlimited war . . . It is a difference of degree rather than of kind.

By contrast, if

> . . . we extend our view to wars between world-wide empires, the distinction at once becomes organic. Possessions which lie overseas or at the extremities of vast areas of imperfectly settled territory are in an entirely different category from those limited objects which Clausewitz contemplated. History shows that they can never have the political importance of objects which are organically part of the European system, and it shows further that *they can be isolated by naval action sufficiently to set up the conditions of true limited war* . . . For our ideas of true limited objects, therefore, we must leave the continental theatres and turn to mixed or maritime wars.[16]

From all this he concluded that by early twentieth-century standards

> limited war is only permanently possible to island Powers or between Powers which are separated by sea, and then only when the Power desiring limited war is able to command the sea to such a degree as to be able not only to isolate the distant object, but also to render impossible the invasion of his home territory.

This then was the key for Corbett to understand England's past success in contests involving much stronger adversaries, in which she managed to gain and maintain the 'command of the sea'.

. . . Except in the light of Clausewitz's doctrine the full meaning of Bacon's famous aphorism is not revealed: 'This much is certain . . . he that commands the sea is at great liberty and may take as much or as little of the war as he will, whereas those that be strongest by land are many times nevertheless in great straits.'[17]

From this Corbett concluded that what Clausewitz called 'war limited by contingent' – i.e., limited interference by one power in an unlimited war conducted between two or more other powers – was Britain's strength. 'During the eighteenth century there had been a large number of cases of war actually limited by contingent – that is, cases where a country not having a vital interest in the object [of a war] made war by furnishing the chief belligerent with an auxiliary force.'[18] '[W]hat may be called the British or maritime form [of war] is in fact the application of the limited method to the unlimited form, as ancillary to the larger operations of our allies – a method which has usually been open to us because the control of the sea has enabled us to select a theatre in effect truly limited.'[19]

Corbett raised the question 'as to whether it is not sometimes legitimate and even correct to aim directly at the ulterior object of the war.' Despite the teaching of Clausewitz and Jomini, most contemporary strategists seemed to think there was only one answer.

Von der Goltz, for instance, is particularly emphatic in asserting that the overthrow of the enemy must always be the object in modern war. He lays it down as 'the first principle of modern warfare,' that 'the immediate objective against which all our efforts must be directed is the hostile main army.' Similarly Prince Kraft has the maxim that 'the first aim should be to overcome the enemy's army. Everything else, the occupation of the country, &c., only comes in the second line.'[20]

British warfare, however, had tended to go for the limited form of war, by contributing limited means to continental allies with the overall unlimited objective of overthrowing the common enemy.[21]

The object of naval warfare must always be directly or indirectly to secure the command of the sea or to prevent the enemy from securing it . . . One of the commonest sources of error in naval speculation . . . is the very general assumption that if one belligerent loses the command of the sea it passes at once to the other belligerent . . . The most common situation in naval war is that neither side has the command . . . [22]

What in fact matters is that the sea should be kept open as a means of communication. 'By denying an enemy this means of passage we check the movement of his national life at sea in the same kind of way that we check it on land by occupying his territory . . . Command of the sea, [however], means nothing but the control of maritime communications, and not, as in land warfare, the conquest of territory. The difference is fundamental.'[23]

Corbett identified two fallacies:

One is, that you can avoid attack by depriving yourself of the power of offence and resting on defence alone, and the other . . . that war consists entirely of battles between armies or fleets. It ignores the fundamental fact that battles are only the means of enabling you to do that which really brings war to an end – that is to exert pressure on the citizens and their collective life. 'After shattering the hostile main army,' says von der Goltz, 'we still have the forcing of a peace as a separate and, in certain circumstances, a more difficult task . . . to make the enemy's country feel the burdens of war with such weight that the desire for peace will prevail.'[24]

Corbett singled out for criticism three main ideas of 'military lore': (1) concentration of force ('the idea of overthrowing the enemy's main strength by bringing to bear upon it the utmost accumulation of weight and energy within your means'); (2) 'that strategy is mainly a question of definite lines of communication'; (3) concentration of effort: concentration on the 'force you wish to overthrow without regard to ulterior objects.'

For the first, Corbett argued that concentration of force was a different matter on the sea because the enemy can simply 'remove his fleet from the board altogether' – one had to work hard to force him into an encounter. As for lines of communication, Corbett argued that the enemy can circumvent the direct route as there are no or few obstacles at sea to prevent him from taking a slightly longer route. Finally, concentration of effort, Corbett thought, should take into consideration that fleets do not just exist to fight decisive battles but also to protect one's own commerce.[25]

In sum, Corbett developed an original theoretical structure which draws on Clausewitz's realistic writing but is self-confident enough to depart from Clausewitz's views where his theories are inapplicable to the sea. A century on, Corbett's views of the role of navies in 'force projection' (manpower, missiles and aircraft) are if anything even more apposite than they were at the time of his writing. In the early twenty-first century, naval battles with navy pitched against navy are hard to imagine, and the naval battles of the two world wars have retreated far into history. By contrast, Corbett's applied Clausewitzianism still stands up.

Corbett is thus in a much more original and conscious way a disciple of

Clausewitz, an epithet which has also, though less justifiably, been claimed for his main intellectual competitor, Alfred Thayer Mahan. While Mahan may have reached similar conclusions to Clausewitz about the value of historical events as a database from which to draw theories, his most recent biographer admits that Mahan probably first read Clausewitz only after he had completed his great *Influence of Sea Power* series. In his advocacy of decisive battles, Mahan may sound at times like Clausewitz the Idealist, but the link lies more in the common study of Napoleonic warfare, which Mahan encountered through the writings of Jomini.[26]

Clausewitz on the 'Small War' or Guerrilla

Another subject area in which Clausewitzian thought was taken further by one of his disciples is that of the 'small war' or the *guerrilla*. Unlike naval warfare, this is a subject on which Clausewitz did write himself, although not all the material he put down is found in *On War*. He wrote not only about 'small war' (in the French tradition of the '*petite guerre*' and the Spanish '*guerrilla*'), but also the people's war, which could be a very intense form of all-out war, as it was during the French Revolutionary Wars and the Wars of Liberation against Napoleon. The terms used by Clausewitz – sometimes interchangeably – are 'the arming of the people' (*Volksbewaffnung*), 'people's war' (*Volkskrieg*), and 'small war'. Today we would say 'low-intensity conflict', in the sense of the definition provided by the US military in 1992:

> [a] political-military confrontation between contending states or groups below conventional war and above the routine, peaceful competition among states. It frequently involves protracted struggles of competing principles and ideologies. Low intensity conflict ranges from subversion to the use of armed force.[27]

In Clausewitz's times, there were several manifestations of the people's war. There was the *levée en masse* of 1793 in France (which was the virtually total mobilisation of the French people for the wars of the French Revolution). Then there was the fighting in the Vendée (French Revolutionaries against Royalists, in which the Revolutionaries massacred large numbers of civilians), and the Spanish *guerrilla* against Napoleon beginning in 1808 from which this form of war took its name (this war was small in respect of the individual units fighting on the Spanish side, but very large and intensive in respect of the proportion of the entire population that contributed to the war effort). These were followed by the Tirolean uprising of 1809, the Russian partisan attacks on the *grande armée* as it was moving through Russian territory in 1812, which Clausewitz had witnessed himself. The clustering of these events led Clausewitz to believe that this form of war was a phenomenon of his own times.[28]

In 1810–11 Clausewitz lectured on 'the small war' at the *Allgemeine Kriegsschule* in Berlin. He defined the 'small war' as a conflict involving anything between 20 and 400 men, mostly irregulars, that is not part of a larger battle. His lectures mainly covered specific military aspects of the small war, without giving any consideration to political and social factors, incentives and motivations. They were very dry and technical, oriented towards tactics, and used many historical examples.[29]

There is an overlap between the small war analysed thus by Clausewitz and the people's war: they can be identical, as in the case of the Spanish *guerrilla* against Napoleon, in which the whole population seemed to have taken up arms against the French invaders, even if the units confronting them were small and avoided pitched battles. But 'small war' can also refer to geographically and technically very limited resistance to regular forces by just a few bands of rebels, probably with the support or connivance of the local population. A people's war invariably denotes the mobilisation of a particularly large number overall, as in the French Revolution's *levée en masse* or the anti-Napoleonic 'Wars of Liberation', or the *Volkssturm* ('people's storm') organised by the National Socialist regime in Germany at the end of the Second World War, movements that might include very young and very old men, and even women, as in the case of Tito's Partisans in the Yugoslav civil war of 1940–45.

If it takes on this last form, the people's war has elements of total war à la Ludendorff. When Prussia was occupied by the French, Clausewitz wrote in his 'Confession Memorandum' of 1812:

> The war of the present is a war *of all against all* [my italics]. It is not the king who fights another king, not an army another, but a people fights another and the people includes king and army. War will hardly change this character again, and it would truly not be desirable that the old bloody and yet boring game of chess of the soldiers' battle would ever come back. But I do not mean by that that the people's uprising in masses [i.e., the *levée en masse*], that we have now seen twice in big examples (France and Spain) will henceforth be the only way in which peoples will wage war against each other, Heaven protect us! That phenomenon is particular to the present with its fateful hours . . . But while there may be future centuries in which none of the peoples is forced to take recourse to the last desperate measure of the people's uprising, we can still say that in these centuries war will be regarded as the business of the nation, and it will be conducted in this spirit.[30]

In *On War*, Clausewitz did not deal with 'small war', but with the 'arming of the people', which he discussed in the context of Book VI on defence. Unlike in his 'Confession Memorandum', he here deliberately abstained from any judgement as to whether the arming of the people was to be welcomed or not.[31] He defined the *Landwehr* ('territorial army') as

> an extraordinary, voluntary contribution of the whole mass of the people to the war, with their bodily strength, their riches and their attitude. The more [the *Landwehr*] differs from this, the more the result will be a standing army in all but name, and the more it will have the advantages of a standing army, while lacking the advantages of a real *Landwehr*, which is a great reservoir of strength, with much wider reach, much less defined, much easier to augment through spirit and attitude.

And like the French strategist Guibert before him, Clausewitz recognised that such a territorial army, and more still the actual arming of the common people (*Volksbewaffnung* or *Landsturm*, 'land storm') is automatically a defensive, and not an offensive tool.[32] He noted the domestic revolutionary potential of the mobilisation of the people, which critics feared 'because they regard it as a revolutionary tool, a legally proclaimed state of anarchy, which is as dangerous to the social order within as for the enemy outside', but Clausewitz added that this problem need not concern him and his readers. He conceded that critics could also claim that the effect of arming people did not merit the effort invested. Be that as it may, Clausewitz regarded the people's war in general as a consequence of the transformation of war that Europe had seen since the French Revolutionary Wars, in which war had transcended its earlier limitations. Once one side had overthrown the barriers that restricted warfare under the *ancien régime*, the other side had to follow or be defeated.[33]

Clausewitz emphasised how many examples of history had proved 'what an enormous factor in the equation of the forces of state, war and army are the heart and the mind of the nation'.[34]

> By its very nature, such scattered resistance will not lend itself to major actions, closely compressed in time and space. Its effect is like that of the process of evaporation: it depends on how much surface is exposed. The greater the surface and the area of contact between it and the enemy forces, the thinner the latter have to be spread, the greater the effect of a general uprising. Like smouldering embers, it consumes the basic foundations of the enemy forces . . . The conditions under which alone a general uprising can be effective, are as follows:
>
> 1. that the war must be fought in the interior of the country,
> 2. that it must not be decided by a single cataclysmic battle,

3. that the theatre of war must extend over a fairly large space,
4. that the people's character must be suited to that type of war,
5. that the country must be rugged and inaccessible, because of mountains or forests, marshes or the nature of the cultivated areas.

Whether the population is large or small is not decisive, as [in a people's war] it is hardly people who are lacking. Whether the population is poor or rich is not particularly relevant either or should not be, but it is difficult to ignore that a class of human beings who are poor and used to hard work and deprivations tend to show themselves as more martial and hardy.[35]

Elsewhere he opined that armed forces with a popular spirit, including enthusiasm, fanatical determination to win, and faith in the cause, can succeed particularly in mountain warfare, where the armed forces break up into small units or even single fighters, all fending for themselves. Mountains are thus the best areas from which to organise a people's war.[36] This is an eerily perceptive prediction of the success of the Yugoslav partisans in pinning down German and Italian divisions in the Second World War, and of the performance of the Greeks in their civil wars of the twentieth century.

Clausewitz made quite detailed prescriptions for the use of the *guerrilla* (which he continued to call 'people's war'). The armed peasants should avoid attacking the centre of gravity of the enemy's occupation forces, but should stage small attacks on their fringes, where they are weakest. They might thus attack an isolated enemy contingent as it was crossing a mountain range or a dense forest, a bridge or a narrow passage. They should rise up in remote provinces; by acting in many different places, the armed people should generally undercut the morale of the enemy's armed forces. They should never try to constitute a standing army but in general remain dispersed, even though they might concentrate their forces on a particular point for the purposes of a surprise attack on the enemy. Clausewitz expressed this rather poetically:

The people's war, like a nebulous and cloud-like being, should never condense itself into a resistant body, or else the enemy will direct an appropriate force against its core, will destroy it and take many prisoners; if that happens, [the people's] courage will flag, everybody thinks the main issue is decided, all further efforts in vain, and the people will let the weapons drop from their hands. Nevertheless, it will be necessary that this fog concentrates on certain points to become a more dense mass and threatening clouds, from which a strong stroke of lightning can erupt from time to time. These points are mainly to be found on the wings of the inimical theatre of war.[37]

The armed peasants should therefore complement the country's own standing army, and that should join up with it from time to time to attack garrisons and harass larger contingents of the occupation forces. The standing forces, or what is left of them under an occupation, should inspire and help organise the armed peasants (*Landsturm*). Yet the regular forces should not be distributed all over the country to form the cadres of the peasant forces, lest they be destroyed together, and lest they become a burden for the logistics of the populations in the provinces.[38] The *Landsturm* itself should not be thought capable of holding out against very determined, forceful occupation forces: it could achieve its best successes in places where it could not be crushed in one fell swoop. 'The strategic plan for defence can include the participation of the arming of the people in two . . . ways: either as the last recourse after the battle is lost or as a natural support, before the decisive battle takes place. The latter presupposes the retreat into the interior of the country.'[39]

Clausewitz laid the foundations of our thinking on asymmetric warfare, by considering on several occasions in *On War* what the relationship is between a stronger and a weaker adversary. He realised that while the best way to victory is unquestionably to have the larger armies and to defeat a smaller or weaker enemy army utterly in one main battle, other factors can favour the smaller or weaker power. Apart from morale, this could be a greater stamina and patience, so that a larger enemy might not be prepared to invest the same amount of time in a particular conflict as the weaker force. This could result from a difference in political importance of the conflict for either of the adversaries,[40] and explains why in the twentieth century France ultimately failed to hold on to Algeria, and why the superpowers USA and USSR failed in Vietnam and Afghanistan respectively.

A number of twentieth-century thinkers on the *guerrilla* found inspiration in Clausewitz's writing. In 1918, when the Russian government concluded the peace of Brest-Litovsk with Germany, Lenin took recourse to Clausewitz to justify his decision: 'If the forces are clearly weak, then the most important means of defence is the withdrawal into the interior of the country.' And, he added, anybody who doubted the verity of this concept 'could read it up in old Clausewitz, one of the greatest authors on war and on the entire teaching on this subject.' Later Soviet official historiography explained this decision thus:

In the period of the October Revolution, Lenin taught the Bolshevik Party how to attack fearlessly and resolutely, if the necessary conditions are present. In the period of the peace of Brest Lenin taught the Party how to stage an orderly withdrawal at a moment when the enemy's forces obviously surpass ours, in order to prepare a new attack against the enemy with great energy. History has proved the utter correctness of Lenin's line of argument.[41]

Also inspired by Clausewitz was Colonel T. E. Lawrence (1888–1935), better known as Lawrence of Arabia, who had read *On War* while an undergraduate at Oxford.[42] In *The Seven Pillars of Wisdom*, Lawrence praised Clausewitz's work as far superior to any other and as having subconsciously inspired him in his own thinking. He was troubled by Clausewitz's enthusiasm in his idealistic phase for the main battle, convinced that the Arabs would not be able to bear the large casualties which such a battle would produce, but took comfort from Clausewitz the Realist's concession that each war and each period had its own peculiar conditions.[43] Lawrence's own enactment of the *guerrilla*, in which he organised Arab freedom fighters against the Ottoman Empire's *Askeriler*, was a very successful form of the 'small war'. It tied down large numbers of regular forces, without, however, bringing about a decision in the overall war, the decisive military encounters eventually taking place in Europe, not in the Middle East.[44]

In his *Modern Strategy*, Colin Gray points out that, in principle, small wars, terrorism, etc., are also a continuation of politics by other means, and that Clausewitz's basic analytical approach ('war is a continuation . . .') can apply even in the absence of the state actors he emphasised so much.[45] (This also counters Martin van Creveld's criticism of 'trinitarian' warfare theories which we discussed in Chapter 3.) Nevertheless, in the *guerrilla* in particular, in the absence of formalised planning practices which we associate with states, action is often taken on the basis of what seems like a good idea at the time, and van Creveld and Gray rightly argue that we reach the limits of Clausewitzian analysis if we assume that every actor holds a clear concept of the purpose of a war and carefully plans the use of force as a function of it.

Mao and Clausewitz

It is arguably one of the Communist disciples of Clausewitz who took his thinking on the people's war to new heights. For Lenin, there were two sorts of war, just or progressive wars and unjust or reactionary wars. The only just wars to him were wars of national liberation or civil wars if they were fought by one class to overthrow the oppression by another. 'National wars by colonies and semi-colonies in the era of imperialism', to Lenin's mind, were 'not only possible and likely, but indeed inevitable', as he wrote in his treatise 'About the Junius Brochure'.[46]

Just as Lenin justified the withdrawal of the Communist forces into

Russia's interior at the peace of Brest-Litovsk in 1918 by referring to Clausewitz, Mao treated the subject of 'Luring the enemy into the interior of the country' in pronounced Clausewitzian terms. He wrote,

> No military theoretician or military practitioner of the past has ever disputed that this is the course of action which a weak army must adopt when faced with a strong adversary, in the initial stages of a war. A foreign military expert [*sic*] once said that in the strategic defensive one should in general initially avoid decisive battles if they would occur under unfavourable circumstances, and seek them only when favourable circumstances have been established. This is entirely correct, and we have nothing to add to it.[47]

And while Clausewitz wrote, 'A quick, forceful transition to the attack – the flashing sword of the revenge – is the most brilliant moment of the defence', Mao (reflecting the probably slightly more ambiguous wording of the Chinese translation) wrote of the counter-offensive as 'the most fascinating and dynamic stage of the defence'.[48]

For Mao, all revolutionary war had to mobilise the people: 'Only if one mobilises the masses can one wage war.'[49] But like Clausewitz, Mao saw the mobilised masses as complementary to regular forces, and it is here that he developed Clausewitz's ideas significantly further, conceptualising different phases of operations in which the war effort would gradually shift from resistance of the mobilised people to those of regular people's forces. These regular forces, in the concrete case of the Chinese Red Army, were indispensable for the final victory, in Mao's thinking; a final, decisive campaign or battle was necessary, and could be fought and won by regular forces alone. The strength of the Red Army lay in its discipline, but also in the support it received from the broad masses through the people's self-defence forces and the militia.

In the words of the scholar Zhang Yuan-Lin, what Mao meant by *guerrilla* was a form of war in which the militia, partisan units or special units from the regular army should fight small enemy units in close connection with the population as a whole, without taking any particular direction or holding a particular front line.

> The general task for the partisans in such *guerrilla* consists of beating the enemy under the cover and with the support of the popular masses, by mainly using surprise attacks. The partisan forces should above all seize areas, entrench themselves and enlarge the areas, while in regular warfare large numbers of regular forces are deployed and the main forms of the conduct of war are the wars of movement and of defending fixed positions.[50]

Mao's teaching bore many resemblances to Clausewitz's five points quoted above (see pages 135–6). He emphatically agreed that the Chinese people's war against Japan should be conducted within the territory of China, which leant itself particularly well to the 'retreat into the interior of the country' on account of its huge dimensions. He also held that the people's war could not be decided by one battle but through a lengthy war of resistance, and that partisan forces should deal out blows against the occupying army in as many different areas as possible, widely scattered throughout China's land mass. He thought the political motivation and thus education of the people's army crucial, and agreed that the partisans should make best possible use of rugged areas of the countryside as their retreats, from where they should make sorties to seize villages and later towns.

A difference lay in Clausewitz's and Mao's views of strong points and bases. Clausewitz did not believe that the partisans should seek to establish them, while Mao declared that 'The lengthiness and duress of the war make it impossible to pursue partisan warfare in the enemy's *hinterland* without [making use of] base areas.' Such base areas he described as

> strategic bases, with the help of which the partisan units fulfil their strategic tasks and reach their aim of preserving their own forces and to grow while annihilating the enemy or chasing him away. Without such strategic places we shall not have the support for the conclusion of all strategic tasks and for the conclusion of the war aims . . . Without base areas partisan warfare cannot last and cannot unfold.[51]

Interestingly, Stalin forced the Greek Communists in the Second Greek Civil War (1946–9) to adopt Mao's doctrine of the need to conquer and hold base areas, which was entirely inappropriate to the much smaller area of land in which the Greek partisans operated; this injunction on the part of Stalin is generally taken to have condemned the Greek Communist campaign to defeat. In 1956, Mao thus wisely warned Latin-American revolutionaries not to apply his base-area doctrine unthinkingly to what might be very different circumstances in their own countries:

> The experience of the Chinese Revolution, i.e., the constitution of bases in the countryside, the encirclement of towns via the villages and finally the seizing of the towns, may well not be easily applicable to many of your countries, but they may serve as inspirations for your own reflections. I want to advise you not to copy the Chinese experience mechanically. The experiences of a different country should always be regarded as an inspiration for further thought, never as dogma.[52]

When in 1975 the West German Chancellor Helmut Schmidt visited Beijing, Mao told him of his admiration for Clausewitz, adding:

Marx, Engels and Lenin have interpreted his famous dictum as though war were nothing special, but merely the continuation of politics with other means. I, by contrast, prefer to read Clausewitz's tenet as a lesson for the military, i.e., even in war the political leadership has to have primacy, and not, as Ludendorff thought, for example, the military leadership. I draw the conclusion that war is only one of many options available to the political leadership. One must never see war as the only option.[53]

As the criticism levelled against him by Martin van Creveld has shown, Clausewitz's contribution to the understanding of the *guerrilla* is widely underestimated, and he is often associated only with 'regular' or 'classical' inter-state war. For instance, the West German author Sebastian Haffner, commenting on the many 'limited' wars which took place during the Cold War, pointed out that Mao's view of war (as aiming at the annihilation and ousting of the enemy rather than 'classical' warfare between governments, which Haffner qualified as 'Clausewitzian'), had been confirmed by reality. Haffner thus, surprisingly, excluded the French Revolutionary Wars from the category of classical, Clausewitzian wars as Haffner defined them; Haffner thus seemed to be writing about the state-to-state warfare known to the *ancien régime* and the wars waged in Europe in the second half of the nineteenth century. Haffner equated Mao's view of war with revolutionary war, people's war, in which the entire population was mobilised. 'Conventional war [of the Clausewitzian sort] is always concerned with one government trying to get another to do what the latter does not wish to do. Apparently that is no longer possible today with [modern] means of war. More recent wars are concerned with the attempts by one regime to abolish another and to replace it.' In this context, the five principles of war which used to rule conventional war are invalidated:

1. Classic European wars were fought with disciplined, professional armies, whose political views did not matter. In the recent wars, which Haffner calls 'total *guerrilla*' à la Mao, discipline and obedience still play a role, but certainly a lesser one: *guerrilla* forces act from conviction, not only as a function of obedience.
2. In classic wars, there was a clear division between armed forces and non-combatants. In the 'total *guerrilla*' these two groups are deliberately mixed. *Guerrilleros* find refuge in the anonymity of the masses of the civilian population who support them.
3. Classic European wars saw hostilities carried into the country of the

enemy. *Guerrilla* warfare, by contrast, tends to be confined to the country from which the *guerrilleros* hail.

4. European military and political leaders aimed to keep classic European wars short. The strength of a *guerrilla* movement, by contrast, partly lies in preventing the end of a war, as long as the victory would go to the stronger regular forces of the enemy. (Mao did, of course, advocate seeking a decisive end to the war if the likelihood was that the *guerrilleros* would win.)[54]

5. Concomitantly, European military and political leaders in 'classical' wars sought a quick decision, a decisive battle. *Guerrilleros*, by contrast, seek to avoid a decisive battle which the stronger regular forces would be more likely to win.

This, so Haffner explained, is why the *guerrilla* need not fear the two factors which made classic war impossible in Europe during the Cold War: its escalation through the total involvement of the people and through modern weapons technology. Once again ignoring Clausewitz's writing about *guerrilla* and *Landsturm*, Haffner summed this up as the supposed triumph of Mao's over Clausewitz's logic of warfare.[55] As we have seen, however, Mao's further developments of Clausewitz's thoughts make this a highly unjust verdict.

More important because of its practical consequences is the influence Clausewitz had, probably via Mao, on other Communist thinkers. Interestingly, one of Mao's Latin-American interlocutors, Ernesto 'Che' Guevara, in his writings on 'Tactics and Strategy of the Latin American Revolution' of 1962, began his deliberations by going back to Clausewitz's rather limited definitions of strategy and tactics, but without building specifically on Clausewitz's work on the 'small war' or on the 'people's war'.[56] And the North Vietnamese Communist thinker Tru'ông-Chinh, writing during the Vietnam War, included in his book, *The Resistance Will Win*, a chapter on military issues in which he repeatedly refers to the Clausewitzian nexus between war and politics. He elaborated that no military success could be achieved if it was not inspired by the right political purpose, and in turn in many cases politics could not succeed without appropriate application of the military tool.[57] While there is thus clear evidence of the influence of Clausewitz on *guerrilla* thinkers who stand in the Marxist-Leninist tradition, none went as far as Mao in elaborating and then applying a coherent teaching.

7

Clausewitz in the Nuclear Age

SOVIET STRATEGY: CLAUSEWITZ AND THE INEVITABILITY OF WAR

We have seen, then, how Clausewitz became part of the Marxist-Leninist canon. By the mid-twentieth century, *On War* had been published five times in the Soviet Union.[1] The Sovietologist Christopher Donnelly even called the Soviets 'Heirs of Clausewitz'.[2]

Yet at the beginning of the nuclear age, Clausewitz's standing in the Soviet Union reached its lowest point, because Stalin saw him as part of the inimical tradition of Prusso-German militarism which had twice in the twentieth century brought death and destruction to Russia. In 1944, the Soviet military press, at the behest of Stalin, severely criticised Clausewitz's findings.[3] In July 1945, Colonel Meshcherjakov published an article in *Voennaja Mysl'*, the confidential interior organ of the Soviet Ministry of Defence, in which Clausewitz, and Lenin's admiration for him, were criticised, Clausewitz having completely failed to understand the class nature of warfare. In response to this article, in January 1946 the Soviet military historian Colonel E. A. Razin wrote a letter to Stalin himself, asking his supreme commander whether it was Lenin or Meshcherjakov who should in future be regarded as the supreme authority on Clausewitz, whom Lenin had, after all, called one of the most remarkable, the greatest and most profound military authors. Razin himself thought that Meshcherjakov's article had little effect but to disorientate the officers and generals of the Red Army and thus to harm it.

On 23 February 1946, Stalin replied to Razin in an open letter. He exhorted Razin to think critically of all German strategists, from Clausewitz and Moltke to Keitel, on the basis of whose teaching Germany had, twice in thirty years, unleashed bloody conflicts, and yet had twice been defeated. Stalin wrote that Lenin had only praised Clausewitz on a political, not a military level, and had never praised Clausewitz for any aspect of his military teaching. When Lenin quoted Clausewitz at the peace of Brest-Litovsk, this was because Lenin needed an articulation of the rationale for retreat rather than attack. Stalin added:

> What must be noted in particular about Clausewitz is that he is, of course, obsolete as a military authority. Strictly speaking, Clausewitz was the representative of the hand-tool [manufacturing] period of warfare. But we are now in the *machine* age of warfare. The machine age undoubtedly demands new military ideologists. It is ridiculous to take lessons from Clausewitz now ... We do not regard Marx's theory as something completed and untouchable; we are convinced, on the contrary, that it had merely laid the cornerstone of that science, which Socialists *must* move further in all directions, unless they want to be left behind by life ...[4]

Indeed, Stalin went on to criticise Engels directly as somebody who had also written in the manufacturing period of warfare, and whose writing was thus hardly applicable to the present age.

After this public rebuke, Razin remained silent on the topic until after the debunking of Stalin at the Twentieth Congress of the Communist Party of the Soviet Union. Other Soviet literature treated Clausewitz with caution or disdain. Only grudgingly was he once referred to indirectly as the father of the German *Blitzkrieg* strategy, which had something in common with Soviet concepts of deep operations dating from the 1930s. Otherwise, no praise was found for him.[5] Indeed, from the 1950s until the 1970s, several Soviet scholars expressed their annoyance with the Western assertion that Lenin was Clausewitz's disciple.[6] In 1951, L. M. Leshinsky wrote on 'The Bankruptcy of the Military Ideology of the German Imperialists', which again followed Stalin's dismissal of Clausewitz.[7] Clausewitz had crucially played down the impact of new inventions in technology on the conduct of war: 'The least number of innovations in the realm of warfare are due to new inventions or new departures in ideas, while most are due to new social constellations and situations.'[8] This is one of the points on which Clausewitz contradicted himself: he himself gave the example of the effect of fire power in the Middle Ages compared to Clausewitz's own time.[9] But this was not picked up by Soviet strategists and theoreticians. By extrapolation, the supporters of Clausewitz in the USSR, who gained ground again after the end of the Stalinist era, argued that the advent of nuclear weapons could not change the essential nature of war as a (rational) continuation of politics.

On 12 March 1954, G. M. Malenkov triggered a debate on nuclear war when he claimed that such a war would entail the destruction of human civilisation, including world Communism, and was thus unacceptable.[10]

Thereafter, there was a lull in references to Lenin's thesis on the inevitability of war.[11]

Two years later, the Eighteenth Congress of the CPSU opened the door to a comprehensive dismissal of Stalin's teaching on a host of issues. Then, in a speech of 27 January 1957, Mao Zedong praised the intellectual achievements of Marx, Engels and Lenin in taking on board classical German philosophy, classical British political economics and French Utopian Socialism, while pointing to the shortcomings of Stalin:

> In this respect Stalin was not so good. For example, during his period, classical German idealist philosophy was called a reaction by the German aristocracy to the French Revolution. This appreciation completely negates the classical German idealistic philosophy. Stalin negated German military science. He said it no longer had any value and Clausewitz's books no longer needed to be read, because the Germans had been defeated after all.[12]

Mao went on to say that, in the eyes of those such as Stalin, 'War is war, and peace . . . peace; the two are mutually exclusive and are in no way connected. [According to them] war cannot be transformed into peace and peace not into war. Lenin [by contrast] quotes from Clausewitz: "war is the mere continuation of politics with other means". Fighting in times of peace is politics, war is also politics, even if it uses special means.'[13]

In this new climate, Colonel Razin felt entitled to return to his charge and to rehabilitate Clausewitz. In 1958 he wrote that the great Prussian had 'influenced the development of military science and the military practice', and had crucially recognised that war was a social phenomenon. To reject his teaching would mean to 'leave his heritage of military theory and history entirely to the reactionary forces'. Instead, he exhorted his countrymen to study Clausewitz's work carefully and thoroughly.[14]

The debate continued. In 1961, General Talensky wrote that war in a military sense was no longer a rational tool of politics, but then, one year later, restated the orthodox line that any future world war would be decided by socio-economic factors, and that Socialism, being the basis of the more advanced socio-economic order, would necessarily win.[15] In his speech to the Supreme Soviet of the USSR of 12 December 1962 on 'The Current Situation and the Foreign Policy of the Soviet Union', Khrushchev himself expressed doubts about whether his country could win a nuclear war.[16] Paradoxically, this made Khrushchev a supporter of Stalin's doubts about the continuing relevance of Clausewitz's teaching on politics and its military tool.

In the general codification of Soviet strategy known by the name of one of its joint authors, V. D. Sokolovskiy, Clausewitz's teaching was fully adopted as a building block of Soviet doctrine, and on this basis the famous formula was put down: 'It is well known that the essential nature of war as a continuation of politics does not change with changing technology and armament.' This of course meant that nuclear weapons had not changed the nature of warfare, which continued to be seen in orthodox Soviet doctrine as a rational tool of politics.[17] Interestingly, it echoed Mao's first reaction to the use of atomic weapons: he had long argued that weapons were less important than human beings, as it was the latter who used the weapons. In 1946, Mao had told the American journalist Anna Louise Streng that 'the atomic bomb is a paper tiger which the USA tries to use to frighten people. It looks terrible, but in reality it is not. Of course the atomic bomb is a weapon of mass destruction. But the people decide how a war ends, not one or two new sorts of weapons.' On another occasion, he referred to the bomb as a 'paper tiger'.[18]

This gave rise to a fierce debate. In 1962, General Talensky restated boldly that 'the time had passed when politics could "play" with war and determine its scale and the forms of struggle' because of war's tendency towards the extreme, which Clausewitz had rightly recognised.[19] That year, the Cuban missile crisis rattled people all over the world. On 24 September 1963, the political commentator Boris Dimitriev wrote wryly in *Izvestija*, 'war can only be a continuation of folly'. Colonel P. Trifonenkov must have earned brownie points from the orthodox side when he retorted in *Krasnaya Zvezda*, on 30 October, that 'the thesis of war as the continuation of politics can never be questioned by a Marxist-Leninist.'[20] And on 11 December, Marshal Sergej Birjusow, Chief of the Soviet General Staff, forcefully reconfirmed the orthodox position in *Izvestija*.[21] In January 1964, Major General N. Sushko and Major Kondratkov challenged this doctrine: they argued that nuclear war, 'because of its clear character of destruction and annihilation, has ceased to be a reliable means to achieve political aims, no matter what those are.' Thus 'war had become an exceptionally dangerous and risky tool of politics.' Sushko and Kondratkov, like Talensky, became the objects of fierce criticism.[22] Amongst others, V. Tsvetkov leapt to the defence of Clausewitz and his writings.[23] Nevertheless, Kondratkov stuck to his guns and frequently repeated his criticism of the Clausewitzian formula in the nuclear age.[24]

A typical reiteration of the orthodox position, dating from 1965, reads as follows:

> The statement that the nuclear missile war has ceased to be a tool of politics is only correct in as far as it has become impossible to resolve the problems and contradictions in the world with the help of war . . . In our times there is no single issue which could justify the use of nuclear weapons. But this does not mean that war has changed its socio-

economic and political nature ... [Nuclear war] would be the
continuation, expression, tool and result of the criminal policies of
imperialism in the pursuit of its aims.[25]

In Sokolovskiy we find that

> the essence of war as a continuation of the politics by means of armed
> violence and the specific nature of war appear today more distinctly than
> in the past, and modern means of violence acquire ever-increasing
> importance ... It should be emphasized that ... the Leninist concept of
> war as a continuation of class politics by forcible means and the concept
> of war as armed conflict in the name of definite political aims remains in
> force even in the present era.[26]

At the Twenty-third Congress of the CPSU in 1966, Army General
Yepichev, the Chief of the Political Central Administration of the Soviet
Armed Forces, confirmed this orthodox tenet: nuclear weapons had not
changed the inevitability of the victory of the Socialist camp in any future
war.[27] In 1969, General Yepichev repeated his assertion:

> As is well known, certain ideologues of Imperialism and politicians and
> military men of the capitalist states seek to prove that Lenin's formula of
> war as continuation of politics with violent means is 'outdated' and
> cannot be applied to a war with nuclear weapons, that such a war
> supposedly cannot have a class political content any longer and can no
> longer be the continuation of politics of States or different classes. The
> purpose of this contention is to mislead the popular masses concerning
> the political class character, the true aims of a possible war and its
> consequences, and to cover up the rôle of the aggressive policies of
> Imperialism in the preparation and unleashing of a war. The classical
> definition which Lenin gave of the nature of war [however] is the
> methodological basis for the scientific logical analysis of the socio-
> political nature and special characteristics of a possible nuclear war. A
> third world war, if the Imperialists succeed in unleashing it in spite of
> everything, will be the decisive clash of the classes of the two
> contrasting social systems. As far as the Imperialist states are concerned,
> this war will be the continuation of the criminal, reactionary and
> aggressive policies of Imperialism. As far as the Soviet Union and the
> countries of the Socialist community are concerned, it will be the
> continuation of the revolutionary policy for freedom and independence
> of the Socialist states as well as for the securing of the construction of
> Socialism and Communism, and will be the just and rightful resistance
> against aggression.[28]

E. Rybkin, Marshal Kulikov and others echoed the theme in the following years: pacifists were criticised as defeatists who were playing into the hands of the imperialists, weakening the anti-imperialist front, if they denied that nuclear war could be waged with the result of an ultimate victory for Socialism.[29] Clausewitz was thus greatly favoured by the orthodox hawks, who after Khrushchev's departure clearly surpassed their critics both in numbers and influence.

Indeed, in the early 1970s, military thinkers like V. Y. Savkin placed Clausewitz on 'the pinnacle of German bourgeois thought': while failing to understand the class nature of warfare, he did a great service to the advancement of theories of war by linking it with politics.[30] And in a treatment of the philosophical heritage of Lenin published in 1972, A. Milovidov and others emphasised again that nuclear war, like any other war, would be the continuation of class warfare, and that political war aims would take into account the possibilities provided by these new technical means.[31]

The third edition of the *Great Soviet Encyclopaedia* approvingly noted Clausewitz as having 'made the first attempt to develop a general theory of war not limited to special questions of strategy and tactics', who famously 'defined war as a continuation of politics by violent means', even though 'he did not understand the class nature of war and reduced it to foreign policy'.[32] One of the 'creators of the art of war', the 'Prussian general's propositions are still valid. The classics of Marxism-Leninism had a high regard for his contributions. A central place in Clausewitz's scholarly achievement is occupied by his proposition about the relation of war and politics: "war is the continuation of politics by other means," and politics contains in a hidden form the main outlines of a future war. "Marxists have always rightly regarded this thesis as the theoretical basis of views on the significance of any war" [Lenin] . . . Clausewitz correctly asserted that "every epoch has its own wars" and that changes in the art of warfare are caused by "new social conditions and relations".'[33]

The debate on the nexus between war and politics in the nuclear age flared up again in the 1980s, in an almost identical rerun of the 1960s. In 1982, during the Euromissile Crisis, General G. V. Sredin of the Red Army wrote:

> The Marxist-Leninist premise that war is a continuation of politics with military means remains true even in the context of fundamental changes in military matters. The thesis of certain bourgeois ideologues that nuclear-charged weapons have removed war from the realm of politics, indeed that nuclear war cannot be controlled by politics, that it is no longer a tool of politics and no longer its continuation, is wrong in theory and is politically reactionary.[34]

In the following year, however, the *Philosophical-Encyclopaedic Diction-ary* published an article by Alexander E. Bovin, political commentator of the *Izvestija*, who argued that while war continued to be the extension of the politics of social classes, it could no longer rationally serve politics as such an instrument, as socio-political and military-technological factors had objectively reduced the likelihood of a global nuclear war.

In 1985, General M. A. Gareyev, Assistant Chief of the Soviet General Staff and the main author of an official biography of Frunze, reiterated the orthodox line that, even today, war could be a logical continuation of politics, and a retaliatory strike with nuclear weapons was a justifiable measure in a war. To negate the Clausewitzian formula meant to cause disorientation in society, and to mask the aggressive policies of imperial-ism. Gareyev wrote: 'It is inadmissible, along with reactionary elements, to disregard Clausewitz merely because he was the military ideologist of the manufacturing period of the conduct of war.'[35]

There is considerable evidence that the Chernobyl disaster that year strengthened the case of the doubters, mainly to be found among civilians. In 1986, Alexander Bovin, writing in the journal *Kommunist*, returned to his charge that Clausewitz was out of date in the nuclear age, as war in pursuit of political objectives no longer made any sense. L. Feoktisov supported his views in the same journal: 'If war is in principle the continuation of politics with other means, these means today serve to make thermonuclear war the continuation of a suicidal and criminal policy which threatens all of mankind.'[36] In 1987, V. Zagladin, a member of the Central Committee of the CPSU, threw his weight behind that of the doubters, stating 'that a nuclear war would not be a continuation of politics.'[37] Anatoli Uktin and Daniil Proektor, both from research institutes on which Mikhail Gorbachov drew so heavily for advice, wrote that 'The Clausewitz formula has lost all its meaning today', and 'The clever "teacher of war", Carl von Clausewitz, is out of date in Europe.'[38] Finally, Gorbachov himself, in his book *Perestroika*, wrote:

> Clausewitz's dictum that war is a continuation of politics with other means, a classic in his own time, is today hopelessly out of date. It should go back on the shelf in the library . . . Nuclear war cannot be the means to achieve political, economic, ideological or other aims.[39]

On 7 December 1987, Gorbachov settled the issue, telling the UN General Assembly that 'It is obvious . . . that the use or threat of force no longer can or must be an instrument of foreign policy.'[40]

The hawks still fought back. Lieutenant General V. Serebrjannikov argued that, more than ever, war had become 'politicised and technologi-sed',[41] and one of his colleagues, Tabunov, warned against rashly throwing out the Clausewitzian formula.[42] But with the assertion of political control by Gorbachov, the Warsaw Treaty Organisation (WTO) was finally given a defensive strategy, and the way was cleared for the Treaty on the Elimination of Intermediate-range Nuclear Forces (INF Treaty) and finally for the Treaty on the Reduction of Conventional Forces in Europe (CFE). With this challenge to Leninism and its adoption of the Clausewitzian formula, Marxism-Leninism itself was in question, finally resulting in its fall with the end of the WTO in 1990 and of the Soviet Union in 1991.

CLAUSEWITZ AND WESTERN COLD-WAR STRATEGY

While it took the Soviet strategists the entire length of the Cold War to reach the conclusion that nuclear war was hardly a rational continuation of politics, most Western strategists reached this conclusion very quickly. The words of the American strategy historian, Russell Weigley, are typical:

> The atomic explosions at Hiroshima and Nagasaki ended Clausewitz's 'the use of combats' as a viable inclusive definition of strategy. A strategy of annihilation could now be so complete that a use of combats encompassing atomic weapons could no longer serve 'for the object of the War', unless the object of war was to transform the enemy's country into a desert. The rational purposes of statecraft could not be thus served. Furthermore, if the United States should lose the monopoly of atomic weapons it possessed in 1945, 'the use of combats' with atomic weapons would almost certainly destroy not only America's enemies beyond rational purpose but the United States as well.[43]

Raymond Aron agreed that in the nuclear age, warfare could no longer be a reasonable continuation of politics by other means. He asserted that the critics of Clausewitz 'cannot get away from the paradox of our age: that is the possibility of unlimited violence which, without the threat being uttered, restrains effective violence.'[44]

Thus in the West, the question of whether nuclear war constituted a rational continuation of politics was mostly answered in the negative, even though nuclear deterrence continued to be seen by a majority of strategists as a useful hedge against major war. This would lead them either to argue that nuclear weapons had assigned Clausewitz's view of war to the dustheap of history, or to emphasise, on the contrary, that nuclear weapons were tools of a rational policy aimed at the avoidance of major war. The espousal by America and NATO of Mutually Assured Destruction (technically confirmed in the agreement to limit anti-ballistic missile defences in the ABM Treaty of 1972) was interpreted as the recognition of the unacceptability of World War III.[45]

By contrast, the Anglo-American war-fighting school of nuclear strategists, which included Keith Payne and Colin S. Gray, argued essentially that 'victory' had to be the aim of military planning while war was an instrument of policy, and they asserted that military planners had no choice, even in the nuclear age, but to assume that it was.[46] Their influence found its reflection in the Presidential Directive No. 59 of July 1980 on 'countervailing strategy'. As one critic commented, 'Clausewitz thereby becomes a medium for promoting efforts designed to prevail at any level of combat, up to and including a full-scale nuclear exchange.'[47]

Countervailing strategy was never, however, accepted by America's allies in NATO. The official viewpoint of one of the USA's principal allies presented the nuclear options for the alliance thus:

> There is no military automaticity in NATO: politics decides on 'the necessary efforts'; it determines which means are appropriate, and then releases them for use by the military leadership. This is true particularly for nuclear weapons. A large-scale nuclear exchange cannot be a meaningful act of war that serves a political purpose, as it would lead to the extinction of both sides, the aggressor and the defender. It has become unlikely, as long as politics remains coupled with this realisation, in which Clausewitz . . . believed . . . Atomic weapons are above all political weapons which serve the prevention of war. They are necessary as long as one of two opponents owns them . . . [Only a] use of atomic weapons that is limited in numbers, well-calibrated in its effect and carefully targeted, can serve a meaningful political purpose . . . Atomic weapons, in whatever way they are used, thus have a mainly political, and only secondarily a military meaning. More than ever the political purpose has to be the measuring rod not only for the military operation to be chosen but also for the kind of means that are to be employed, and that purpose and means have to correspond. The nuclear age does not contradict Clausewitz, but instead confirms [his teaching] most convincingly. 'For the political intention is the purpose, war is the means, and the means must never be considered without the purpose.'[48]

Western strategists soon embraced the creed of deterrence, i.e., the hope that the fear of nuclear war would suffice to keep the other side from deliberately initiating major war. The Europeans were even more willing to profess their faith in this doctrine. A typical European assertion – that in the nuclear age, war has to be dominated by political considerations more than ever – can be found, for example, in a West German publication of excerpts of Clausewitzian *dicta*.[49] The danger was recognised, however, that nuclear war might result from an accident or a miscalculation, escalating from a minor clash, and Clausewitz was regarded as having laid the basis for the concept of escalation.

Western Cold-War Strategists and Clausewitz's Heritage

Western strategists in the nuclear age tried to draw lessons from the two world wars. Bernard Brodie, for example, called World War I, 'the greatest catastrophe in modern times, [which] may have more lessons for the future than World War II, which was in fact its offspring'.[50] 'World War I was the purposeless war, which no one seemed to know how to prevent and which, once begun, no one seemed to know how to stop.'[51] Could this catastrophe repeat itself in the nuclear age?

Some argued that in Clausewitz's concept of absolute war as formulated in Book I of *On War*, where it is described as the abstract idea of a discharge of force without any friction, made him the 'prophet of the apocalypse', and the 'incredulous historian of an apocalyptic future', of an all-out nuclear World War III.[52] More frequently, Clausewitz was evoked in a negative sense: how could war still be a *rational* continuation of politics by other means if the consequence of war could not but be disproportionate to any political aim? Even in 1945, Brodie wrote:

[T]he first and most vital step in any American security program for the age of atomic bombs is to take measures to guarantee to ourselves in case of attack the possibility of retaliation in kind. [I am] not for the moment concerned about who will *win* the next war in which atomic bombs are used. Thus far the chief purpose of our military establishment has been to win wars. From now on its chief purpose must be to avert them. It can have almost no other useful purpose.[53]

And he concluded:

On the simple Clausewitzian premise ... – that war must have a reasonable political objective with which the military operations must be reasonably consonant – we have to work back from the assumption that 'general war' with thermonuclear weapons must never be permitted to begin, however much we find it necessary to make physical preparations as though it might begin. Working back from that premise is far from easy, and ... the idea of large-scale conventional war is simply no solution. There are requirements for a new diplomacy, the beginnings of which are in fact appearing ... We must ... take up [Ivan] Bloch's old question of the *possibility* of that war which must not be, and which would today and in the future really produce that national suicide . . . which we now know applies to our time as it did not to his.[54]

Brodie argued repeatedly 'that nuclear weapons do by their very existence in large numbers make obsolete the use and hence need for conventional force on anything like the scale of either world war.'[55] Brodie was merely the first of many to have drawn this conclusion from the use of atomic bombs against the hapless cities of Hiroshima and Nagasaki. As we have seen, and as the West German scholar Werner Gembruch pointed out, while Clausewitz believed that weapons technology could serve to distinguish different periods of warfare, he on the whole gave little attention to its importance in warfare. Gembruch was one of many who raised the question

> whether, in view of the art of war as a task of politics . . . given modern scientific-technical developments not only of missiles and atomic weapons which have opened up the possibility of a perfect concentration of destructive power in space and time, war can still be a means of resolving political issues . . . Does technology kill war, because it renders war increasingly lethal? Does the total power of annihilation . . . make at least 'major war' impossible?[56]

If the answer was 'no', then there might still be the hope of preventing it from reaching its extreme form through 'escalation dominance'. To explain this concept, one must first turn to Clausewitz's contribution to the concept of escalation.

Clausewitz and Escalation

As early as 1810–12, in his lectures for the Crown Prince, Clausewitz laid out a cost-benefit calculation for war: 'Naturally one always seeks to have the likelihood of success on one's side in war, by counting on one's physical or moral superiority. But this is not always given; often one has to go against the odds if there is no better option left.'[57]

In Book I of *On War* he wrote:

> Humane souls might tend to think that there are artful ways of disarming or shattering the enemy, without causing too many wounds, and that this is the true aim of the art of war . . . Such errors, which arise from kindness, are the worst . . . He who uses force ruthlessly, careless of how much blood is shed, will gain preponderance if his adversary does not do the same . . . [But if both act in this way], both

sides will escalate [*steigern*] to the extreme [*zum äußersten*], without there being any limits other than the forces used to check each other . . .

We repeat our tenet: war is an act of force, and there are no limits to its application; thus each side imposes itself upon the other, which leads to an interaction which according to the ideal [nature of war] has to lead to the extreme.[58]

If war can be limited, it can become a process of bargain: 'The aim of warfare is to make the enemy unable to defend himself.' But Clausewitz the Realist argued that this was a theoretical construct, for in reality, this escalation is avoidable:

If the enemy is to do our will, we must place him in a situation which is even more unpleasant than the sacrifice which we demand from him . . . Any change of this situation through further fighting, must lead to an even less advantageous situation [for the enemy], at least in his imagination. The worst situation in which a belligerent may end up is that of complete defencelessness. Thus if the enemy is to be compelled to do our will through the act of war, we either have to make him defenceless in fact or put him in a situation in which he is threatened with defencelessness.[59]

It is this ability to bargain with threats of escalation which stops real wars from escalating to the extreme.[60] In Book I, Chapter 2, we read: 'The smaller the sacrifice we demand from our adversary, the weaker his efforts will be to deny it to us. The weaker his efforts, the less force we need to use ourselves.'[61] Clausewitz proceeded to discuss the

likelihood of success without defeating the enemy's forces, i.e., such undertakings that have direct political repercussions, that are designed in the first place to disrupt the opposing alliance, or to paralyse it, that gain us new allies, that favourably affect political developments, etc. If such undertakings are possible it is obvious that they can greatly improve our prospects and that they can form a much shorter route to our goal than the shattering of the inimical armed forces . . . [62]

[In the extreme case,] an estimate may be enough. In such cases, there will be no fighting: the weaker side will yield at once.[63]

And in Book VIII, Chapter 2, we read:

One does not start a war, or one should reasonably never start a war, without determining what one wants to achieve in this war and by means of this war, the former being the war aim, the latter the purpose. This determining principle will dictate all the directions, the scale of the means and the intensity of energy employed, and it will be reflected even by the smallest detail of the action.[64]

Clausewitz thus emphasised the need to take all possibilities of escalation into account before venturing into a war. He restated this idea in Book VIII, Chapter 3A:

Theory . . . demands that at the beginning of a war, its character and dimensions should be determined on the basis of probability, which results from the political factors and circumstances. The closer these probabilities drive the character of war toward the absolute, the more its dimensions include the full weight of the belligerent states and draw them into its vortex . . . the more necessary it is not to take the first step without considering the last.[65]

This, then, is the Clausewitzian foundation of the concept of escalation. This concept of course made Clausewitz particularly interesting in the nuclear age. If the last escalatory step in such a war would be the use of nuclear weapons, then one had to hesitate to take the first step. But this concept applies equally to any major war, even if nuclear weapons did not exist. It is difficult to believe that either the monarchs and governments who began the First World War, or indeed Hitler or the Japanese government on the eve of the Second, anticipated the 'last step' and its likelihood. And it was major wars of this sort that Clausewitz had experienced himself, and that he was writing about: it does not need weapons of mass destruction to make major wars unbearable, particularly for industrial or post-industrial societies, as Ivan Bloch had argued even before the First World War.

Western Cold War strategic writers applied Clausewitz's ideas on escalation in an attempt to turn the danger of nuclear escalation to the advantage of one's own side in war.

Most famously, Herman Kahn of the Hudson Institute defined 'escalation' as 'an increase in the level of conflict in international crisis situations'. 'In a typical escalation situation,' wrote Kahn

there is likely to be a 'competition in risk-taking' [a term he attributes to Thomas Schelling] or at least resolve, and a matching of local resources, in some form of limited conflict between two sides. Usually, either side could win by increasing its efforts in some way, *provided that the other side did not negate the increase* by increasing its own efforts. Furthermore, in many situations it will be clear that if the increase in effort were not matched and thus resulted in victory, the costs of the increased effort would be low in relation to the benefits of victory. Therefore, the fear that the other side may react, indeed overreact, is most likely to deter escalation, and not the undesirability or cost of the escalation itself. It is because of this that the 'competition in risk-taking' and resolve takes place. . . .

Building thus on the Clausewitzian concept of war as dialectic of two wills, Kahn continued:

> . . . in any escalation, two sets of basic elements are in constant interplay: the political, diplomatic, and military issues surrounding the particular conflict, and the level of violence and provocation at which it is fought. The latter merges with those considerations raised by the possibility of escalation to higher or more extensive levels of violence, including the possibility of a deliberate, provoked, or inadvertent conflict eruption leading directly to central war. Just as there are two basic sets of elements in the escalation situation, so there are two basic classes of strategies that each side can use. One class of strategies makes use of the features of the particular 'agreed battle' that is being waged in order to gain an advantage. The other class uses the risks or threat of escalation and eruption from this agreed battle.[66]

One of Kahn's disciples, Stephen Cimbala, in *Clausewitz and Escalation* argues that there are 'pregame', 'midgame' and 'endgame objectives': 'pregame objectives' are to avoid outbreak of war without surrendering key political aims; 'midgame objectives' concern escalation management; 'endgame objectives' are to end war without too much destruction while protecting or achieving political aims.[67]

We recall Clausewitz's double trinity, as discussed in Chapter 3, with the correlation of violence and hatred with the public at large; of uncertainty and chance with the military command; and of policy decisions with the government. Taking this as a starting point for his analysis, Cimbala argues that, in the nuclear age, uncertainty and chance are problems for government as well as for military leaders. Cimbala is convinced that the war plans of the nuclear powers could not have been implemented in practice. 'The assumption behind much US nuclear strategy has been that preparedness for total war would deter lesser wars.' US and NATO strategy thus espoused a choice of options which fell short of total war and might still lead to war termination without too much destruction.

Far from being able to impose control over US and NATO alliance decision-making as they proposed to do, American and allied leaders would find the immediate effects of even small nuclear wars devastating to the morale of troops and field commanders and paralysing to the persons responsible for top-level political decision-making. Whereas one can infer from Clausewitz, mistakenly in my view, that certain aspects of a damage limitation and counterforce, escalation-dominance nuclear strategy might provide a valid connection between war and policy, this judgement would be incorrect.

By 'escalation dominance', American strategic writers mean that aspect of nuclear strategy which tries to steer clear of all-out escalation to city bombing for as long as possible and which initially uses nuclear strikes in a limited way, to 'bargain' with the opponent. This could be done to limit damage to one's own side by encouraging the enemy to abstain from targeting NATO cities; such encouragement might take the form of strikes against the enemy's (nuclear) forces (known as counterforce strikes), again avoiding what was dearest to him, namely his cities and industries (value-targets).

For Cimbala, Clausewitz's theory of friction leads to the suspicion that any nuclear war might easily get out of control.[68] Cimbala differentiates between different sorts of friction – simple friction, compound friction ('the interaction of people, plans and technologies in unexpected ways') and complex friction ('overlaps with compound friction . . . insufficient or erroneous intelligence appreciations; faulty assumptions about the opponent's strategies, operational art and tactics; and communications of other failures of information transmittal and retrieval'),[69] and stresses the importance of uncertainty, but also that of deception and intelligence, which in the nuclear age is more important than it was in Clausewitz's age.[70] From this Cimbala deduces that it is hardly possible any longer to differentiate between offensive and defensive in the nuclear age. Deterrence (with a negative aim) is matched by compellence (with a positive aim) – both can exist in the nuclear age.[71] He continued:

Clausewitz's philosophical depth and military experience made it possible for him to discuss the theory and practice of war in meaningful and original ways. His many important insights include a strong appreciation of the problem of escalation. However, there were limits to what Clausewitz was able to accomplish with regard to the clarification of strategy in general, and with respect to the problem of escalation in particular. First, Clausewitz was limited by his own background . . . to a state-centered system of war and deterrence based on the European balance of power system and on domestic principles of dynastic legitimacy. Second, he lived and wrote before the development of many technologies which have made war very different in degree, and in the case of nuclear weapons technology, in kind. Third, he offered no middle range theory to bridge the gap between his abstract concepts of war in theory and his detailed observations of actual war. These limitations have important implications for what we can infer from *On War* about nuclear deterrence and escalation.[72]

Cimbala notes that since the beginning of the nuclear age, strategic surprise was a very real possibility, while in Clausewitz's age it was not.[73] Nevertheless,

> Clausewitz identifies three aspects of the problem of escalation: the reaction between emotion and other decision-making criteria; the interaction of leaders' opposed wills and objectives; and third, the *reciprocal* effects of each side's escalation on the other's assumption about *subsequent* escalation. This perception of the problem of escalation was remarkably far sighted, and has a number of implications for the management of nuclear crises and for the control of escalation in war today, including, if necessary, nuclear war . . .

Generally, then,

> Clausewitz's understanding of the relationship between war and policy, and of the forces which cause conflict to expand or to contract, offers three general categories of findings of interest to students of escalation and control in nuclear strategy.
> First, crossing the threshold from peace to war, from coercion to the actual use of force, is more significant than the first use of nuclear weapons for nuclear armed states . . . Clausewitz repeatedly calls attention to the uniqueness of the wartime, compared to the peacetime, environment . . .
> Second, US and other analysts have contended over whether conflict spiral models or deterrence models better explain the growth of a smaller war into a larger, or the deterioration of a crisis into war. Undoubtedly both models are necessary for a complete explanation of escalation. Nonetheless, there are critical differences of emphasis. Deterrence models emphasize the communication of credible threats to inflict unacceptable punishment as the key to stability. Spiral models, on the other hand, emphasize the risk that the credibility of deterrence threats can be provocative, instead of dissuasive, of the behavior the

threatener is trying to prevent. Some threats can be too credible. The expectation by either the US or the Soviet leadership during a crisis, for example, that the other side had a credible first strike capability could raise the first side's incentives for pre-emption instead of reducing them . . .

Support from Clausewitz can be inferred for both the spiral and the deterrence models of escalation control.[74]

While the impact of Clausewitz's thinking is once again clear, not everybody was happy with these Clausewitzian attempts to think through the unthinkable. Anatol Rapoport criticised the enthusiasm of Herman Kahn, Raymond Aron and other 'Neo-Clausewitzians', who in Rapoport's views misunderstood the master in some important ways. They

attempted to reduce nuclear war and deterrence to a matter of calculable rationality, susceptible to such mathematical technique as game theory. And . . . it was precisely to this intellectualisation of war, this reduction of a bloody tragedy to a mathematical problem, this elimination of all moral and political content from the complex equation, that Clausewitz himself was objecting. Kahn and his colleagues in their studies achieved the remarkable result of ignoring all three elements in the Clausewitzian trinity: popular passion, the risks and uncertainties of the military environment, and the political purpose for which the war was fought. Their calculations bore no relation to war as mankind has known it throughout history.[75]

Policy Prescription or Contemplative Theory?

Ironically, it was Colin Gray – who a decade later was to become a policy adviser of the Reagan administration – who in 1971 argued that civilian strategists should seek only the pursuit of truth, not of 'policy prescription' or 'the advocacy of viable solutions'. He argued that in the US they had 'fallen between the two extremes', over-impressed as they were 'with the potential transferability of theory to the world of action.'[76] Bernard Brodie took issue with this point, invoking Clausewitz to argue that 'strategic theory is a theory for action'. Brodie thought Gray's attitude was a 'sad retrogression from Clausewitz', and himself claimed that 'strategy is a field where truth is sought in the pursuit of viable solutions. In that respect it is like other branches of politics and like any of the applied sciences, and not at all like pure science, where the function of theory is to describe, organize, and explain and not to prescribe.'[77]

Michael Howard (who together with Peter Paret edited the timely new translation of Clausewitz's *On War* which became so influential in American military thinking after the setback of the Vietnam War) took an active part in the nuclear debate within NATO on the basis of Clausewitzian considerations. He was rightly characterised as an 'owl', whose views were somewhere mid-way between those of the 'doves' (the disarmers and pacifists) and the 'hawks' (including the nuclear-war fighters like Colin Gray). In the latter context, Howard accused some of his colleagues of neglecting the factor of national will and political context, arguing that 'there is no way in which the use of strategic nuclear weapons could be a rational instrument of State policy.' He thus dismissed Colin Gray's call for a 'war-winning capacity' as irrational, arguing that Gray and like-minded strategists advocated it 'not because their masters have any serious political motive for extirpating the societies of their adversaries, but because in a grotesque inversion of logic the means now dictate the ends.' Howard therefore advocated an increase in conventional forces, which in his view could form a meaningful defence capacity, meaningful in a Clausewitzian sense.[78]

But, like Brodie and Howard, Colin Gray in *Modern Strategy*, published at the end of the century, builds heavily on Clausewitz, whose *dicta* crop up throughout the text. Like Clausewitz himself, Gray is caught on the horns of the dilemma of on the one hand insisting, emphatically, on the 'unitary nature of strategy', the 'unity of all historical strategic experience',[79] and of pointing out, on the other, that each conflict is distinct and is influenced by cultural peculiarities as well as by the means available. Moreover, in his chapters on nuclear weapons (which include a section on 'Clausewitz and the bomb'[80]), Gray asserts on the one hand that nuclear weapons have a strategic utility 'when the Clausewitzian language of "the engagement" is interpreted to encompass "deterrence action", which is to say threats and latent menaces at work in the minds of those intended to be deterred,' but on the other hand that these are 'weapons that could not reliably be tamed for good Clausewitzian purpose as tools of high policy' and writes that 'after the mid-1960s, the political leaders and the military professionals of East and West were appalled by the monster of nuclear armament that they had created which could not, *après* Clausewitz, be a *rational* instrument of state policy.'[81] Thus when it comes to nuclear weapons, the question is not *whether* they are instruments of state policy, which they clearly are, but whether they are rational instruments of state policy. Taking it one logical step further, the true problem is that our ideas of rationality are not universal, and are almost certainly not shared by every Dr Strangelove and every Bin Laden in the world. We should be careful to note that if something seems rational to us, it does not mean it would be the priority of other cultures, and other decision-makers.

Limited War and the Western Neo-Clausewitzians

Clausewitz's concept of limited war became popular among strategic thinkers in the nuclear age with the advent of the Korean War (1950–53), which was called by Bernard Brodie the 'first modern limited war'.[82] The beginning of the Cold War had been dominated by the conviction that if it turned hot, it could only result in total (or at least major) war; indeed, when the Korean War broke out, policy analysts throughout NATO were convinced that this was merely the first in a series of moves which would include major war in Europe. When it remained an isolated case, strategic thinkers turned to Clausewitz for an explanation of this phenomenon, and to Clausewitz's realist vocabulary of limited war.

As we recall, the Clausewitzian concept of 'limited war' had been declared obsolete by Colin and others on the eve of the First World War. At the time of the Korean War, however, American defence academics like William Kaufmann, Robert Osgood and Morton Halperin rediscovered Clausewitz's concept and elaborated it. Oddly, however, their entire focus was only on limited war between America and the USSR or China; the 'general rules' of American political sciences concerning limited war in reality applied only to America and its military policies in the very specific Cold War context.

Kaufmann argued that the USA should show willingness to reach a compromise with the Communist adversary if it wanted to avoid the (Clausewitzian) escalation of a crisis to nuclear war. He thought that Europe was an area in which it was unlikely that such a compromise could be reached and a limited war could be waged. The problem with Kaufmann's 'realist' appreciation of the situation was that it did not take into account the ideological incompatibility of Communism and Western-style democracy. His idea of a compromise presupposed that each side could tolerate the continued existence of the other and indeed even the partial realisation of its aims. This was difficult in view of the tension between the two sides arising from their mutually exclusive ideologies, posited on both sides on the need to see the other destroyed. The situation was quite different from the rivalry of eighteenth- and nineteenth-century European princes, each of whom wanted to see the continuation of the social structure that had placed them at the top.[83] Kaufmann rightly argued that the Cold War context and the pursuit of limited war aims required the contesting sides to 'give up the concept of victory in its traditional meaning. We discard the idea of punishing aggression,' but only the Western side gave up that pursuit of victory (and even then not entirely, as America's experiments with countervailing strategy showed. The Warsaw Treaty Organisation clung to the pursuit of victory until 1987.[84]) Kaufmann argued:

Given the weapons systems already available to the nations of both East and West, the limitations which meet this test [of being a limited war] are those that restrict the scope and methods of conflict. Ideally, one might wish to see a war confined as to area, targets, weapons, manpower, time, and tempo; but it seems doubtful that so many limitations could be imposed or maintained simultaneously. Probably the most we can hope and work for are restrictions of area and weapons ... Needless to say, neat, bloodless quadrilles such as the War of the Bavarian Succession are too much to expect. To the extent that limited wars are imaginable at all, they are conceivable, not as neatly ordered and umpired contests, but as rough and ready affairs, varying in their character from place to place and being played out under restrictions suitable to the particular occasion.[85]

Kaufmann reminded his readers that eighteenth-century wars were limited by political, economic, technical and other factors. By contrast, he thought it was very difficult for nuclear powers to wage limited wars because only the cost-benefit calculations could persuade them not to raise the stakes if they got nowhere with limited means. As we have noted, Kaufmann and his colleagues wrote exclusively about limited wars between the USA and the Soviet Union, or the USA and China. In both cases, confronted with a major (nuclear) power, Kaufmann thought it important to avoid the complete annihilation of the enemy's armed forces; at best, the steady attrition of the enemy's forces seemed desirable.[86]

In 1957, Kaufmann's colleague Robert Endicott Osgood published his famous treatise, *Limited War*, which he saw as a *Challenge to American Strategy*. Notwithstanding the fact that the Korean War, like many other bloody conflicts during the Cold War, was to an important extent a civil war, Osgood defined war 'as an organized clash of arms between sovereign states seeking to assert their wills against one another.' 'War is a contest between national wills.' This sentence is not the only echo of Clausewitz. Osgood called war 'the upper extremity of a whole scale of international conflict of ascending intensity and scope'. And he asserted, with explicit reference to Clausewitz's concept of war as part of political intercourse, 'The primacy of politics in war means, simply, that military operations should be conducted so as to achieve concrete, limited, and attainable security objectives, in order that war's destruction and violence may be rationally directed toward legitimate ends of national policy.'[87] Osgood's main concern in writing this treatise was to explain that, contrary to the traditional American way of war with its tendency to deploy massive force and to aim for the utter defeat of the enemy, in limited war 'victory is not an end in itself', and 'the whole conduct of warfare – its strategy, its tactics, its termination – must be governed by the nature of a nation's political objectives and not by independent standards of military glory'. Almost paraphrasing Clausewitz, he noted that war might offer only a limited choice of options to politics, which had to accept compromises in view of these limitations; yet 'the wisdom of such compromises must still be judged by their relation to some superior political objective'.[88]

But, again in a Clausewitzian vein, Osgood also noted that war as an instrument of politics had a dangerous tendency to develop to the extreme, to escalate beyond the original political intentions of the actors involved.[89] His advice was thus:

1. Statesmen should scrupulously limit the controlling political objectives of war and clearly communicate the limited nature of these objectives to the enemy . . .
2. Statesmen should make every effort to maintain an active diplomatic intercourse toward the end of terminating the war by a negotiated settlement on the basis of limited objectives . . .
3. Statesmen should try to restrict the physical dimensions of war as stringently as compatible with the attainment of the objectives at stake, since the opportunities for the political control of war . . . tend to decrease as the dimensions of war increase and tend to increase as the dimensions of war decrease.[90]

In conclusion, Osgood explained,

the principal justification of limited war lies in the fact that it maximizes the opportunities for the effective use of military force as a rational instrument of national policy. In accordance with this rationale, limited war would be equally desirable if nuclear weapons had never been invented. However, the existence of these and other weapons of mass destruction clearly adds great urgency to limitation. Before nations possessed nuclear weapons, they might gain worthwhile objectives consonant with the sacrifices of war even in a war fought with their total resources. But now the stupendous destruction accompanying all-out nuclear war makes it hard to conceive of such a war serving any rational purpose except the continued existence of the nation as a political unit – and, perhaps, to salvage the remnants of civilization – in the midst of the wreckage. Only by carefully limiting the dimensions of warfare can nations minimize the risk of war becoming an intolerable disaster.[91]

Two decades later, Osgood partially revised his theory of limited war, acknowledging that most of the wars that had occurred since 1945 had entailed an element of intra-state warfare, and was thus not the warfare between sovereign states he had described in his earlier treatise, even if an element of superpower intervention was mostly present.[92] From the perspective of the world's great powers, he described two strands of limited-war theories that had become common currency:

> One strand, inspired by the concepts of Clausewitz and propounded by Western political scientists and defense specialists, has sought to make force, in both war and deterrence, an effective instrument of containment against the Soviet Union, China, and the international Communist parties aligned with them. The other strand, inspired by Mao Tse-tung and Third-World nationalism and propounded by revolutionary nationalists, has sought to use guerrilla warfare to abolish Western colonialism and hegemony and establish new nations ostensibly dedicated to social justice.[93]

Both strands of limited-war theory aimed to avoid escalation to general war, but aimed to maximise the political utility of force below that level of violence. Consequently, said Osgood, they had to be 'fought for ends far short of the complete subordination of one state's will to another's, using means that involve far less than the total military resources of the belligerents and leave the civilian life and the armed forces of the belligerents largely intact.'[94]

From a Western perspective, said Osgood, limited wars were fought in full view of the danger that local wars might become vehicles of Communist expansion and might prove uncontainable, and that they might expose the ineffectiveness of the American nuclear deterrent in limited circumstances, thus raising the spectre in Europe that the American deterrent might not cover the USA's transatlantic allies either. Thus in the early 1950s, 'limited-war strategy thrived on opposition to the prevailing Eisenhower-Dulles strategy of increased reliance on nuclear deterrence'.[95] Again Osgood explained:

> Although the principal stimulus of limited-war strategy was the perceived imperative of military containment in the nuclear age, the underlying rationale, as expounded by academic analysts and public leaders, transcended the cold war. It rested on the Clausewitzian principle that armed force must serve national policy and therefore, lest it follows its own rules to the physical limits of violence, must be restrained and controlled in order to serve specific political objectives of the state by the use of means proportionate and appropriate to the political stakes and circumstances. The purpose of war, according to this principle, could not be simply to apply maximum force toward the military defeat of the adversary; rather, it must be to employ force skilfully along a continuous spectrum – from diplomacy, to crises short

of war, to an overt clash of arms – in order to exert the desired effect upon the adversary's will.[96]

And again using Clausewitzian images, including commerce and the contest of wills, Osgood wrote:

> The conduct of limited war came to be seen as part of a general 'strategy of conflict' in which adversaries would bargain with each other through the medium of graduated military responses, within the boundaries of contrived mutual restraints, in order to achieve a negotiated settlement short of mutual destruction. The 'escalation' of war – that is, the graduated increase of its scope and intensity – although originally feared as an uncontrollable danger, came to be regarded as a controllable and reversible process by which adversaries would test each other's will and nerve in order to resolve their conflict at a cost reasonably related to the issues at stake.[97]

Indeed, the Cold War abounded in instances of limited war. Focusing only on the USA and the USSR the Harvard scholar Morton Halperin produced in 1963 a list of instances when they had clashed directly or by proxy: Cuba in 1962, the Taiwan Straits in 1955 and 1958, Korea in 1950–53, China in 1946–9, Lebanon in 1958, Greece in 1946–9, and Berlin in 1948 and 1958–62. On each occasion, force was used or threatened, but all examples came under the heading of 'local war', as homelands of the USA and USSR did not come under attack (which would be 'central war'). All these crises were limited both in geography and in intensity. Halperin imagined that central war could also be limited, in view of the desire on both sides to prevent escalation to thermonuclear war, and differentiated between an 'explosion' of a situation, or a gradual and deliberate 'expansion' of hostilities, both of which amount to escalation. He also professed scepticism at the possibility of all-out victory in central war, but emphasised that limited military success is possible in a limited war, and limited war does present options other than total escalation and surrender.[98] Because of the existence of thermonuclear weapons, the two superpowers had a vital interest in avoiding either 'explosion' or 'expansion'.[99]

Like most American strategists, Halperin defined 'limited war' as: 'a military encounter in which the Soviet Union and the United States see each other on opposing sides and in which the effort of each falls short of the attempt to use all of its power to destroy the other.' His study was thus concerned only with Cold War East-West confrontations which has US and Soviet support, and is thus of little relevance to a more general study of limited wars in a post-Cold War setting. Yet he claimed generally to be writing about 'locally limited war, or what will be called here simply "local war". Much of the military-strategic literature has used the terms "limited war" and "local war" synonymously. However, during the past few years it has become clear that central war, that is, a war involving attacks on the homelands of the two major powers, may also be a limited war.' While most of Halperin's study was focused on 'locally limited war', it also included a study of the options of limiting a 'central war'.[100] Halperin deliberately excluded from his studies wars and crises which could not be squeezed into the US–Soviet pattern of confrontation.[101]

While Halperin's terminology was Clausewitzian in part, Thomas Schelling explicitly put himself in the tradition of the Prussian. Schelling explained that even though the technical differences between a tactical nuclear weapon and conventional weapons might be so small as to be insignificant, because of the international revulsion against nuclear weapons, it would not be the same to use them in a war that one seeks to limit.

> What makes atomic weapons different is a powerful tradition that they *are* different . . . Traditions and conventions are not simply an analogy for limits in war, or a curious aspect of them; tradition or precedent or convention is the essence of the limits. The fundamental characteristic of any limit in a limited war is the psychic, intellectual, or social characteristic of being mutually recognized by both sides as having some kind of authority, the authority deriving mainly from the sheer perception of mutual acknowledgement, of a 'tacit bargain'.[102]

Schelling used the term 'strategy' (as in 'strategy of conflict') not as something that was 'concerned with the efficient *application* of force but with the *exploitation of potential force*.'[103] Thus he saw any crisis situation as a bargaining process, in which various measures are threatened or hinted at by both sides. He greatly developed Clausewitz's theme of war as a form of commerce, using game theory. Brodie, too, saw the value of this approach, eplaining that 'The Korean War proved anew that great-power rivals occasionally prefer to test each other's strength and resolution with limited rather than unlimited commitments to violence, and it demonstrated also some of the major constraints necessary to keep a war limited.'[104]

But from the experience of Korea, Brodie, Osgood and others concluded that 'What is a limited war to us, limited in terms of emotional as well as material commitments, may be total war to our opponents and to one or more of our allies; thus, their demonstration of "resolve" may well exceed ours.' Osgood noted also that the same war could be limited and quite unlimited, depending on the perspective of the belligerents: thus the Vietnam War was hardly a limited war from either a North or South

Vietnamese perspective.[105] The Clausewitzian paradigm needed refinement: there was not merely the need to distinguish between wars of different intensity on a sliding scale of limited to unlimited, but the same war could be different things to the different contestants. Moreover, again in the words of Brodie, 'The United States is indeed a very great military power, but when it fights in a limited manner it automatically cuts itself down to a size that the opponent may be able to cope with, even if only temporarily, thus raising our costs and prolonging the war.' If pressure equals counter-pressure, then the result is not movement but a stalemate.[106] Korea showed this in part, but Vietnam was an even better example of this problem with Clausewitz's definition.

A 'Seminar on Capabilities and Techniques of American Armament for Limited War', organised by the American Ordnance Association in New York on 4 December 1957, produced the following definition: 'A limited war is a war fought to achieve a limited objective. In the achievement of this objective a nation may be expected to plan to expend a limited amount of its national resources; and in carrying out the war it may be expected to plan to hold the war to a limited geographical area.'[107] Brodie commented:

> We cannot have limited war without settling for limited objectives, which in practice is likely to mean a negotiated peace based on compromise. Clausewitz's classic definition, that the object of war is to impose one's will on the enemy, must be modified, at least for any opponent who has a substantial nuclear capability behind him. Against such an opponent one's terms must be modest enough to permit him to accept them, without his being pushed by desperation into rejecting both those terms and the limitations of the fighting. This principle, if consistently pursued, should dispose of the hackneyed argument that limited war is impossible because the losing side will always be constrained to reject limitations rather than accept defeat. We have already admitted that keeping war limited will be difficult. We must be clear, however, that the curtailing of our taste for unequivocal victory is one of the prices we pay to keep the physical violence, and thus the costs of penalties, from going beyond the level of the tolerable. It is not the other way round.[108]

Taking a similar approach, Raymond Aron showed how Clausewitz's breakdown of strategy as a function of political purpose, military means and political will, can also be applied to understand the evolution of the limited wars of the twentieth century. For instance, one can explain why the differing strength of the political will of France on the one hand and the Vietnamese Communists on the other led to the ultimate defeat of France in Indochina, as was true in the case of the French in Algeria and the Americans in Vietnam, despite the superior military means of France and America respectively. Clausewitz's framework of analysis thus lends itself to the explanation of the outcome of asymmetrical conflicts, provided all three levels – purpose, means and will – are taken into consideration. Admittedly Clausewitz merely provided the corner stones for the development of such a framework of interpretation, but as these include the important factor of morale, the resulting framework is very useful even for the interpretation of conflicts which Clausewitz could not have foreseen.[109]

The Clausewitzian Critique of the Vietnam War

Retrospectively, the Vietnam War presented a very particular example of American limited war – and a particularly unsuccessful one. Again, Clausewitz was seen to provide a framework for a better understanding of what had happened and what had gone wrong. In 1982, Colonel Harry G. Summers published 'A Critical Analysis of the Vietnam War', which he called *On Strategy*. 'It might seem incongruous', he wrote, that most of the analysis in this book was 'drawn from a 150-year-old source – Clausewitz's *On War*. But . . . this is the most modern source available.' Unlike in economics or political science, the one great classic, *On War*, had not in his view been superseded by any more recent literature: 'In military science . . . *On War* is still the seminal work.' He agreed with Bernard Brodie that 'most of the contemporary books [do] not, as Clausewitz does, have much to say of relevance to the Vietnam war.'[110]

So what were the reasons for American failure in Vietnam? Summers drew up a long list, each time quoting passages from *On War* to explain where mistakes had been made.

One reason, he believed, was the confusion over what Clausewitz called the mere 'preparation for war, and war proper', the latter requiring the intelligent application of the fighting forces, equipment, etc., in the war itself – the employment of the armed forces to secure the objectives of policy. Summers underlined that the political objectives themselves were flawed, as they did not meet Clausewitz's criterion for what constitutes appropriate ends of strategy: the choice of 'objectives that will finally lead to peace'.[111]

A second reason for American failure in Summers' view lay in the excessively submissive and uncritical relationship between the military on the one hand and the civilian decision-makers in the US government. In his introduction he wrote,

> Carl von Clausewitz warned that the military must always be
> subordinate to political policy. This was no news to Americans, for we

had written that very premise into our Constitution fifty years earlier. But Clausewitz went on to say that this subordination rests on the assumption that 'policy knows the instrument it means to use.'[112]

His contention was that the civilian decision-makers in the US government had not known enough about the possibilities and limits of the armed forces, and that the military leaders had not made enough of an effort to explain these to their civilian masters, instead accepting orders which fighting forces are generally not designed or equipped to execute. This unquestioning obedience, in Summers' view, 'grew out of our neglect of military strategy in the post-World War II nuclear era. Almost all of the professional literature on military strategy was written by *civilian* analysts – political scientists from the academic world and systems analysts from the Defense community.' Like Bernard Brodie in his book *War and Politics*, Summers drew attention to the 'lack of professional military strategic thought'. Army officers felt that it was not up to them to make strategy but to concentrate on procurement, training, resource allocation, while actual strategy was made at a higher level and was budget-driven.[113]

Meanwhile the 'pseudo-economic ideology' of systems analysis dominated the Pentagon. Secretary of Defense Robert McNamara's civilian systems analysts in the Planning, Programming and Budgeting System (PPBS), whose involvement in the planning and execution of the US involvement in Vietnam was crucial, handled only the *preparation* for war well. They were incapable of giving the right orders for the *conduct* of the war. In this context Summers pointed to Clausewitz's dictum that 'the economic approach to strategy "stood in about the same relationship to combat as the craft of the swordsmith to the art of fencing".'[114] Nevertheless, leading systems analysts contended that it was not the job of the military to analyse or design strategies, that this was, in other words, too serious a business to be left to the military, who should content themselves with conducting military operations. They held that '[m]odern-day strategy and force planning has become largely an analytical process . . . [and] civilians are often better trained in modern analytical techniques.' The problem with such an approach, as Summers observed, was that it seemed to presuppose a thoroughly predictable, static opponent, that it based analysis on predictions for the number of tanks produced in the US and the number produced in North Vietnam, that it assumed that attrition rates could be safely deduced from known factors such as fire power and the quality of the equipment used. As Clausewitz had observed, however, the essential problem with such calculations was that 'in war the will is directed at an animate object that reacts.'[115] Summers added the 'bitter little story that made the rounds during the closing days of the Vietnam war':

When the Nixon Administration took over in 1969 all the data on North Vietnam and on the United States was fed into a Pentagon computer – population, gross national product, manufacturing capability, number of tanks, ships, and aircraft, size of the armed forces, and the like. The computer was then asked, 'When will we win?' It took only moments to give the answer: 'You won in 1964!'[116]

Again, Summers used Clausewitz's words to explain this error in the calculations of the systems analysts:

They aim at fixed values; but in war everything is uncertain, and calculations have to be made with variable quantities. They direct the inquiry exclusively towards physical quantities, whereas all military action is intertwined with psychological forces and effects. They consider only the unilateral action, whereas war consists of a continuous interaction of opposites.[117]

In turn, the military were baffled by the tasks they were given, but remained unquestioningly obedient to their civilian masters, thus adopting a jargon and a strategy they did not fully understand or believe in. Henry Kissinger commented retrospectively that the military 'permitted themselves to be co-opted too readily', inculcated as they were by the commitment to civilian supremacy. Meanwhile a 'new breed of military officer emerged: men who had learned the new jargon, who could present the systems analysis arguments so much in vogue, more articulate than the older generation and more skilful in bureaucratic maneuvering. On some levels it eased civilian-military relationships', but on another level it stifled fundamental criticism of the civilian strategy which was making inappropriate use of the military tool.[118]

A third mistake Summers identified was the failure of the US government to mobilise the American people's support for the war effort, pointing at Clausewitz's analysis of the difference between the eighteenth-century cabinet wars and the wars of the French Revolution and Napoleon, which in his view had been so successful because war had become 'the business of the people'. As we recall, this was the central tenet of Clausewitz's secondary trinity, in which strategy is influenced by the dynamics between government, army and people. Summers developed that into the tenet that government, army and people had to share the same aims and convictions for a war effort to be irresistible.[119] Vietnam was a failure as Americans had seen Vietnam as 'Johnson's war' or 'Nixon's war', not a nation-wide issue.[120] This failure to mobilise popular support lay in the fact that the key military figures, 'with World War II fresh in their minds ... equated mobilizing national will with total war, and they

believed total war unthinkable in a nuclear age.' They disregarded the fact
that the US had fought several limited wars in the nineteenth century, and
that each had involved the mobilisation of public support.[121] In the absence
of public support, the US was much less committed to the war than were
its opponents, and Summers aptly quoted Clausewitz once more:

> Not every war need be fought until one side collapses. When the
> motives and tensions of war are slight we can imagine that the very
> faintest prospect of defeat might be enough to cause one side to yield. If
> from the very start the other side feels that this is probable, it will
> obviously concentrate on bringing about this probability rather than
> take the long way round and totally defeat the enemy.[122]

In Summers' view, the Johnson administration did not even do what was
necessary to get full support from Congress, the people's representatives.
The Gulf of Tonking Resolution passed by Congress in 1964 gave the US
President sweeping powers by authorising him to 'take all necessary
measures to repel an armed attack against the forces of the United States
and to prevent further aggression.' Johnson interpreted this as a mandate
for all his subsequent commitments of forces to South-east Asia. It was
only in November 1973 during the Nixon presidency that Congress
reclaimed its power to declare war (Article I, Section 8, of the US
Constitution) by passing the War Powers Resolution, limiting the
President's right to deploy military forces to up to a maximum of 90 days,
unless Congress specifically authorised further use of force. Between those
two dates, there was a domestic political debate in the US about the
legitimacy of the armed forces in Vietnam, undermining the morale of
those who served in them – and thus also affecting the second leg of
Clausewitz's 'trinity', the armed forces themselves.

Summers also pointed to a problem which arose from the presentation
of the war. He noted that Clausewitz had warned against excluding 'all
moral factors' and reducing 'everything to a few mathematical formulas'.
In Summers' view, it was courting moral outrage to bureaucratise the
language used by the military to describe their actions. 'We did not kill the
enemy, we "inflicted casualties"; we did not destroy things, we "neutral-
ized targets".' At the same time, the bloody truth behind these
euphemisms (many of which, incidentally, were not unique to the Vietnam
War, let alone new terms) was brought home to millions of television
spectators every evening, and the reaction to the incongruence between the
language of government and military pronouncements and the visible
horror of the war was one of a further erosion of what little public support
there had been. As Summers observed, 'Vietnam was fought in cold blood,
and that was intolerable to the American people.'[123]

With regard to the formulation of strategy, its application in war and the conduct of the war itself, Summers again identified several reasons for America's failure. The way in which the Americans waged the wars in Korea and Vietnam, in the view of Lieutenant Colonel Albert Sidney Britt III, was derived partly from eighteenth-century warfare.[124] Summers showed how Clausewitz's criticism matched the situation of the Vietnam War, in which the US armed forces were more proficient in terms of management than military strategy. In the eighteenth century 'the terms "art of war" and "science of war" were used to designate only the total body of knowledge and skill that was concerned with material factors. The design and use of weapons . . . the internal organization of the army, and the mechanism of its movements constituted the substance of this knowledge and skill.'[125] Thus it was again at the time of the Vietnam War.

Meanwhile the strategy formulated by the civilians failed to formulate war aims which were realistically achievable. Summers accused the Johnson and Nixon administrations of having been overly concerned with the danger of an escalation of the conflict and subsequent Chinese intervention (as had happened in the Korean War) or even Soviet intervention, either of which they feared might lead to nuclear war. Henry Kissinger observed:

> We had entered the Korean War because we were afraid that to fail to do so would produce a much graver danger to Europe in the near future. But then the very reluctance to face an all-out onslaught on Europe severely circumscribed the risks we were prepared to run to prevail in Korea . . .
>
> Ten years later we encountered the same dilemmas in Vietnam. Once more we became involved because we considered the warfare in Indochina the manifestation of a co-ordinated global Communist strategy. Again we sought to limit our risks because the very global challenge of which Indochina seemed to be a part also made Vietnam appear as an unprofitable place for a showdown . . . [126]

This fear of escalation is reflected in the language of the 1954 American *Field Service Regulations* which introduced the concept of 'wars of limited objective'. At the same time, it removed 'victory' as a war aim: 'Victory alone as an aim of war cannot be justified, since in itself victory does not always assure the realization of national objectives.' Harry Summers commented:

> Defining victory only in terms of total victory, rather than more accurately as the attainment of the objectives for which the war is waged, was a strategic mistake [of the Field Service Regulation's

authors]. It not only obscured the fact that we had won a victory in Korea (where the status quo ante was restored . . .), it also went a long way toward guaranteeing a lack of victory in Vietnam.[127]

The *Field Service Regulations* of 1962 replaced the concept of limited war with that of limited means. The manual postulated, 'The essential objective of United States military forces will be to terminate the conflict rapidly and decisively in a manner best calculated to prevent its spread to general (nuclear) war.'[128] Consequently, the US from the beginning declared the limits of its involvement in Vietnam, something which its adversaries could exploit in their own strategy.

Summers concluded that the resulting strategy sought excessively to localise US military action to South Vietnam, with initially only severely circumscribed air action against North Vietnam. The ardent desire to limit the war resulted in its misperception as predominantly a civil war between the South Vietnamese Viet Cong Communists and the forces of the South Vietnamese regime. This had essentially been the nature of the First Indochina war between France and the Viet Minh. The Second Indochina War, or Vietnam War, by contrast, was a long drawn-out campaign on the part of North Vietnam to conquer South Vietnam, which flexibly adapted to the enemy's capability. While the Americans were deployed in South Vietnam, Hanoi avoided set battles and acted primarily through its South Vietnamese fighters, the Viet Cong, in low-intensity operations and the one exceptional high-intensity operation of the Tet Offensive of 1968, when the North Vietnamese regime was ready to sacrifice huge numbers of the Viet Cong. Once the Americans had safely pulled out, the Hanoi regime went over to regular, high-intensity battles, resulting in the conquest of South Vietnam in 1975.

Instead of concentrating on North Vietnam as the enemy's centre of gravity, then, the successive US governments became obsessed with the strategy of 'counterinsurgency', operations limited to the territory of South Vietnam. Revolutionary, people's wars based on Mao Zedong's teaching – purely internal conflicts with an ideological agenda – were seen as the wars of the future, and most intellectual effort was concentrated on trying to find an appropriate *political* riposte to this local form of war. As Sir Robert Thompson, the British adviser to South Vietnamese President Ngo Dinh Diem, pointed out, revolutionary war is 'often confused with guerrilla or partisan warfare . . . The main difference is that guerrilla warfare is designed merely to harass and distract the enemy so that the regular forces can reach a decision in conventional battles. . . . Revolutionary war on the other hand is designed to reach a decisive result on its own.'[129] Applying this definition, Summers argued that Vietnam was clearly a case of *guerrilla*, and the mistake was to treat it mainly as a Revolutionary War.

'South Vietnam faced not only internal insurgency but also outside aggression, and counterinsurgency doctrine could only be part of the answer.'

Yet most definitions of counterinsurgency, and above all those adopted as policy by the US government, involved state-building tasks in South Vietnam – the creation and consolidation of loyalty of the citizens to their government, their inoculation against Communist temptations. Oddly, the US armed forces in South Vietnam were expected to carry out these tasks (much as NATO forces were in the 1990s in Bosnia-Hercegovina and in Kosovo, with equally meagre success). The 1968 *Field Service Regulations* thus read: 'The fundamental purpose of the US military forces is to preserve, restore, or create an environment of order or stability within which the instrumentalities of government can function effectively under a code of laws.'[130] Summers argued that the US Army was ill-equipped for this task, quoting Clausewitz: 'If statesmen look to certain military moves and action to produce effects that are foreign to their nature . . . political decisions influence operations for the worse.'[131] As a former commander of the US forces in South Vietnam and subsequent Chief of Staff commented, the military was 'called upon to perform political, economic and social tasks beyond its capability while at the same time it was limited in its authority to accomplish those military tasks of which it was capable.'[132] In Summers' view, only the South Vietnamese themselves might have succeeded in the state- (or 'nation'-) building tasks which the Americans decided to shoulder in the South, as only they had a credible long-term stake in the country. He quoted Clausewitz: 'One country may support another's cause, but will never take it so seriously as it takes its own. A moderately-sized force will be sent to its help; but if things go wrong the operation is pretty well written off, and one tries to withdraw at the smallest possible cost.'[133]

Meanwhile, for a long time the US did little to hit at the true centre of its opponent's gravity, North Vietnam itself. Actions against the North were limited to air strikes and naval operations. General Maxwell Taylor told the US Senate in 1966 that it was not the US objective to defeat North Vietnam, but merely 'to cause them to mend their ways'. To use Clausewitzian terms again, victory against the principal adversary was thus excluded as an objective.[134] As Summers wrote, 'Instead of orienting on North Vietnam – the source of war – we turned our attention to the symptom – the guerrilla war in the south.' Even Laos and Cambodia, other sources of supplies or transit routes for the Viet Cong, were long treated as sanctuaries not to be attacked, at any rate by US ground forces. Summers conceded that a 'strategic offensive against North Vietnam may not have been politically feasible', but America 'could have taken the tactical offensive to isolate the battlefield' in South Vietnam by cutting off all supplies from outside its territory. 'Since the insurgency itself was a tactical screen masking North Vietnam's real objectives (the conquest of South Vietnam), our counterinsurgency operations could only be tactical, no matter what we called them.' The US, by focusing on the minor centre of gravity and sanctuarising the major centre of gravity, thus failed to adopt a strategy that could lead to peace – the only valid strategy in Clausewitzian terms. 'Ironically, our tactical successes [in South Vietnam] did not prevent our strategic failure and North Vietnam's tactical failures did not prevent their strategic success . . .' North Vietnam's strategy 'caused the United States to deploy against a secondary force' (the Viet Cong in South Vietnam) 'and exhaust itself in the effort. It also caused the Army of South Vietnam to deploy in such a manner that it could not be massed to meet a conventional North Vietnamese cross-border conventional attack.'[135]

This failure in strategic perception and planning was partly due to the lack of unity of command, according to Summers, with the consecutive US administrations never transforming themselves into a war government but continuing with a peacetime decision-making structure. The latter was perfectly adequate for a prolonged Cold War situation, thought Summers – or in Clausewitz's terms, for the preparation for war – but not for the conduct of a real live hot war in South-east Asia. Unity of command was lacking on both the civilian and military levels in Washington, but also in the theatre: the strategic headquarters of the Pacific Command charged with the conduct of operations was located in Honolulu, 5,000 miles from the actual fighting. Summers naturally contrasted this with Clausewitz's injunction that political and military authority should be closely co-ordinated, to the extent that the chief of military operations should sit on the war council.[136]

From Vietnam to the Gulf War

A fresh reading of Clausewitz thus became closely linked with criticism of the American way of war in Vietnam. In the words of Michael Handel, professor at the US Naval War College:

The faculties of the US war colleges, and students who graduated in the late 1970s and early 1980s, gradually (and for some, reluctantly) understood that war and politics could not be separated as was often assumed; that military victory does not automatically guarantee ultimate political victory; that all wars, in particular those that are prolonged, require the political support and consensus of the people in whose name they are waged; that the military is only one of the three elements essential for success in war (the others being the government and the people); and that without a harmonious balance among these three elements, wars cannot be won no matter how just the cause or how great the effort invested. They also discovered the danger of fighting a war with irresolute political backing (which is almost inevitable in any prolonged war given the nature of the American political system). Furthermore, it was realized that gradual escalation and other difficulties imposed by various political constraints prevented the concentration of maximum force that was necessary from the start. Above all, bitter experience had proved that it was 'imperative . . . not to take the first step without considering the last.'[137]

The lessons of the Vietnam War led the Reagan administration to adopt a doctrine named after the US Secretary of Defense, Caspar Weinberger. He formulated six 'tests' which had to be applied before using US forces in combat abroad. Several of these reflected – directly or indirectly – Clausewitz's teaching:

THIRD, if we do decide to commit forces to combat overseas, we should have clearly defined political and military objectives. And we should know precisely how our forces can accomplish those clearly defined objectives. And we should have and send the forces to do just that. As Clausewitz wrote, 'No one starts a war – or rather, no one in his senses ought to do so – without first being clear in his mind what he intends to achieve by that war, and how he intends to conduct it.'

FOURTH, the relationship between our objectives and the forces we have committed – their size, composition and disposition – must be continually reassessed and adjusted if necessary. When they do change, then so must our combat requirements. We must continuously keep as beacon lights before us the basic questions: 'Is this conflict in our national interest?' 'Does our national interest require us to fight, to use force of arms?' If the answer is 'yes', then we *must* win. If the answer is 'no', then we should not be in combat.

FIFTH, before the US commits combat forces abroad there must be some reasonable assurance we will have the support of the American people and their elected representatives in Congress . . . We cannot fight a battle with Congress at home while asking our troops to win a war overseas or, as in the case of Vietnam, in effect, asking our troops not to win, just to be there.[138]

Albeit under another administration, the Weinberger tests or Weinberger Doctrine, as Michael Handel commented, 'was shown to be an outstanding framework for developing an effective strategy tailored to the American domestic political environment.' In the Gulf War of 1990–91, the elder Bush's administration 'demonstrated that it was possible to fulfil all of the necessary political and military conditions; and by doing so, they made the military victory against Iraq possible.' Indeed, it seemed to all sides that in this conflict a fair balance between the constant imposition of political requirements and the absence of excessive political interference ('micro-management') in the military operations was struck.[139] In terms of the implementation of the UN Security Council's injunction to restore the status quo that had existed before Iraq's invasion of Kuwait, the Gulf War was a Clausewitzian victory (it imposed the UN's will on Saddam Hussein, who had to give up Kuwait). A decade later, however, Saddam Hussein was still in power and was still thwarting the International Atomic Energy Authority's verification attempts in pursuit of his nuclear programme, which the UN Security Council had also willed him to give up. Defined in larger terms, the victory of the coalition forces thus seemed ephemeral. US and British air forces were still being used in 2001 in a violent attempt to force Iraq to abide by this second part of the UNSC's resolution. Nevertheless, the Weinberger Doctrine continued to make itself felt in subsequent US security policy, particularly in the operation against Afghanistan's Taliban regime and terrorist leader Bin Laden in the winter of 2001/2002.

To conclude, then, the limited war theory developed by American strategists was entirely US-centric and cannot be applied more generally. But the Weinberger Doctrine, with its Clausewitzian language, contains several principles which could be applied to the security policy of countries other than the USA. It has become a convention that nuclear powers do not use their nuclear weapons in war; still, there is the question of what they would do if they were confronted by an enemy using chemical or biological weapons (leaving aside their self-denying ordinance reiterated in connection with the extension of the Non-Proliferation Treaty in 1995, by which they have engaged themselves not to use nuclear weapons against non-nuclear adversaries). It is also an open question how the non-recognised nuclear powers, above all India and Pakistan, will relate to the taboo of nuclear use. In the Cold War, these two countries have repeatedly come into direct conflict with each other; and they have been the only nuclear powers to have had a direct military clash with each other. It remains to be seen what difference their open nuclear status since 1998 will make to this relationship in the twenty-first century.

At any rate, there is no safety in Colin Gray's assumption that deterrence works in the same way between any 'pair' of nuclear adversaries, and that limited war is the only option open to them. As Cimbala has reminded us, Clausewitz would urge anyone to beware of war's tendency towards escalation, to get out of hand, and of passions aroused by war. And it is worth reminding ourselves that while *ideally* all governments should follow Clausewitz's precept of thinking about the possible last step before venturing on the first, not all governments in Europe or elsewhere have always done so.

8

Clausewitz's Relevance in the Twenty-First Century

At the beginning of the twenty-first century, Clausewitz's concepts and language have deeply penetrated the world's military literature, whether they are correctly or incorrectly used. Whether it is a Russian general or an American political scientist writing about the importance of chance in warfare, whether we read about centres of gravity and what they might be, or about 'Compellence and the Role of Air Power as a Political Instrument',[1] we are dealing with derivatives of Clausewitzian terms and ideas. War as an instrument of politics, war as a contest of wills, limited war and unlimited war, friction, the danger of escalation, all these are Clausewitzian concepts which today are virtually taken for granted in the military manuals and the key literature of the world's leading military powers. As the term 'strategy' is ever more generally and widely used, Clausewitz has even attracted the attention of business-management specialists, albeit in a very unscholarly fashion. Taking *On War* as a quarry for randomly selected and applied *dicta*, Clausewitz now serves as a guide on 'How to Conquer Markets'.[2]

There have long been voices claiming that Clausewitz's main work has become obsolete. We recall that General Colin had declared it so in 1911. Colin's Prussian contemporary, General Walter Reinhardt, on the whole 'considered Clausewitz's theory to be obsolete for a number of reasons, among them the impact of modern firepower . . . the increased importance of logistics; and the blurring of the division between strategy and tactics.'[3] The German General von Kuhn, questioned about the conduct of World War I, told a Reichstag Committee: 'In spite of the high esteem in which the work "On War" is held by every soldier, not everything contained in it is still, after a hundred years, unconditionally applicable.'[4] And as we have seen, Ludendorff wrote in 1935 that 'all theories of Clausewitz have to be thrown overboard', as 'his work belongs to a past period of World History and is mostly out of date. One may even get confused by studying it.'[5] And in 1987, Gorbachov relegated *On War* to the shelves of his library.

Yet we have seen in the previous chapters that other strategists and scholars refer to *On War* as a great work, or, in Bernard Brodie's words, as the 'only truly great book on war'. On balance, we find that *On War* has aged better than Colin's or Bernhardi's or Ludendorff's rival works, not to mention the works of his all-but-forgotten contemporaries, including Jomini.

Problems with Clausewitz

This is not to say that, even leaving aside the problems of interpretation, there are no serious problems with Clausewitz's work. *On War* does not cover a whole host of issues, which Clausewitz either ignored (such as naval strategy) or which have become important since (such as economic warfare, psychological and propaganda warfare, war with weapons of mass destruction and the use of space and information technology for military purposes). Terms like 'global strategy', 'world power', 'superpower', 'world war' are not part of Clausewitz's vocabulary. Nevertheless, scholars and practitioners have identified Clausewitzian *dicta* which they have found applicable to some or all of these areas.[6]

We have noted that Clausewitz was only happy with his revised Book I when he had to cease his work on *On War* in 1830. As a result, there are not merely whole books in *On War* which deal exclusively with 'ideal war' or what Clausewitz called 'absolute' war. The consequence is also that Clausewitz never managed fully to think through some of the implications of his own discovery of the relationship between political aims and the conduct of war.

One such central issue is Clausewitz's idea, simplified in terms of the 'dual nature of war' as he expressed it in his letter to a friend of July 1827 and in parts of Book VIII, in which he correlates revolutionary political aims with absolute war; and limited political aims with limited war. In a very insightful article, Eberhard Kessel dissected this correlation in the light of subsequent history, and found it wanting. For one, Clausewitz's example of a limited war aim, the conquest of a small patch of territory either to keep it or to exchange it for something else in peace negotiations, stemmed from the pre-Revolutionary eighteenth century which he had studied so intensively, when the concept of national territory did not yet exist. Instead, territory belonged to a prince or a dynasty, and was easily passed from the competence of one administrative centre to one geographically far away if it was either inherited by somebody in another branch of the family, given away as a dowry in marriage, or conquered in war (where the conquest was often legitimised by a rivalling dynastic claim to the territory). While the age of absolutism defended the sovereign rights of princes over all their territory, it was equally used to having princes deal with their possessions as they saw fit: their subjects had no right to protest if they were handed over from one jurisdiction to another, from one owner to another.

Clausewitz himself lived on the threshold of a new age, when the subjects of princes increasingly redefined themselves (and were encouraged to redefine themselves) as citizens of a state, members of a nation that inhabited a particular, well-defined 'national' territory. Thus by the time the Franco-Prussian War came, the war aim of the pan-Germans who wanted to found a state containing all Germanophones was to separate German-speaking Alsace from France. What would have been a purely economic and military-strategic issue in the eighteenth century (for example, one might have imagined the Prussians proposing the cession of a part of German territory, or the connivance of Prussia in the conquest of southern Belgium by France, in return for the cession of Alsace and Lorraine), by 1870 had become an issue of *national* pride. The integrity of the *national* territory of France was questioned in one small area, and this was an attack on French sovereignty as a whole. Likewise, a century on, NATO was prepared to risk all-out war to restore the status quo should the Warsaw Treaty Organisation forces attempt to seize even a small area of NATO territory (such as Hamburg and therefore the access to the North Sea); the least territorial conquest would have been seen as an unacceptable loss of the integrity of NATO, representing a defeat not only in real terms but also in terms of the credibility of the alliance as a whole. It is interesting to note how much Clausewitz was still conditioned by the world into which he had been born, even though the new world of nationalism and of 'nation-states' was being conceived in his head as in those of his contemporaries. Politically determined frontiers were still subject to frequent change with the many limited wars of the eighteenth century. But by 1957, when Kessel was writing, any attempt to change frontiers between adversaries was not worth the cost in violence and destruction which it would most likely escalate to. And it is worth adding that the many new state frontiers which came into being at the end of the Cold War through the disintegration of the Soviet Empire invariably followed old administrative demarcations; hardly any attempts were made even after the beginning of the twenty-first century to modify these frontiers.

The second and related point that Kessel made was that since Clausewitz's times, wars even with limited aims (war aims that at any rate did not necessarily count among them the destruction of the enemy's state and the colonisation of his population and territory in the way in which

Napoleon had proceeded) tended increasingly – at any rate in Europe – to be fought as absolute wars. Coming from a German perspective, it is difficult to argue with the 'narrative' of the drift towards increasingly absolute war that Roger Chickering has identified.[7] Under Moltke and Bismarck, Prussia applied unlimited strategies to seize limited territory from Denmark and France, and to force Austria and France to accept that the German princes would in future handle their relations themselves under the aegis of a central government, headed by the Prussian king as the emperor of their aggregate territory, now referred to as 'Germany'. These were all limited aims, as they hardly aimed at the destruction of the political order of other European states; nor were German war aims in the First World War even politically definable. And yet the way the Prussians and Germans fought these wars came ever closer to Clausewitz's ideal war or 'absolute war'. Military considerations predominated, or else the conviction that even a limited war aim would elicit an unlimited response on the part of the enemy. Clausewitz had asked himself in Book VIII whether future wars would always tend towards the absolute and be 'the business of the people', or whether they would once again become the wars of cabinets; the answer, at least up to the Second World War in Europe, was that war remained a business of the people, and as long as it concerned attacks on *national* territory, the people were automatically mobilised and it is difficult to imagine how wars between these nations could remain limited.

Kessel also pointed out that Clausewitz had not pondered the question of what a war would be like in the case of a substantial asymmetry of war aims, i.e., if one side had unlimited war aims and the other had merely limited aims. Surely, in such a configuration, the latter would be forced to step up his efforts to match those of his adversary or face certain defeat. Kessel noted that Clausewitz's Book VIII lacked a chapter on a defensive war where the enemy aimed for the complete annihilation of one's forces and the overthrow of one's government, or where he merely had a limited aim.[8] Kessel came to the conclusion that the conduct of war was determined by military means to a much greater extent than Clausewitz had allowed for, and by the potential of unlimited war. Kessel thought that the very discovery of the concept of absolute war had changed people's consciousness:

> As soon as the idea of absolute war has entered into the consciousness of humanity, it had to affect the choice of aims and the execution of military actions in every instance in an escalatory way. Moreover there is another effect which Clausewitz did not predict, namely that concomitantly, it became increasingly difficult to terminate a war with diplomatic means. This means that today *every* war, even those which

originally are not absolute wars, increasingly tends to become an absolute war, so that the danger of escalation to absolute war is finally contained in every war.[9]

Therefore 'it can be desirable and useful, even if the war aim is limited, to try to attain it with a maximum of force, not least in order to conclude the campaign as quickly as possible.'[10] The generals commanding the coalition in the 1990–91 Gulf War agreed with this sentiment.

Like Martin van Creveld, whose criticism of the 'trinity' we discussed in Chapter 3, some critics faulted Clausewitz for his state-centric approach. Johan Galtung, a West German specialist on international relations, criticised Clausewitz's thinking on war as 'extremely naïf', as it took as its starting point 'horizontal-symmetrical structures of conflict, in which both warring parties are taken as equally autonomous, etc.' Galtung himself thought it necessary to differentiate between external or inter-state wars; internal or intra-state wars, i.e., civil wars; imperialistic wars (within one block or empire); wars of liberation or subversive wars; and international class warfare. (Each of these five categories he subdivided further, in the typical fashion of theoreticians of international relations seeking to contain within their theoretical frameworks all the individual cases they knew, until the subdivisions tended to match at most one or two concrete cases, which they might as well have specified under their own names.)[11]

Another critique of Clausewitz's writing on limited war has come from the strategy expert Jan Willem Honig. In his view, 'Clausewitz never developed a theory of limited war.'[12] In other words, Clausewitz recognises that wars can have limited political purposes, and that in turn this will dampen the military effort made in one or more ways (for example, fewer forces may be made available for the war effort, less effort may be made to mobilise popular support for the campaign). Clausewitz did not, however, dwell on the limited operational aims which should flow from limited political ends. He did not, for example, say that limited political ends determine that only a small battle will be fought, rather than a major, decisive battle resulting in the annihilation of the forces on one side. We must concede that Clausewitz manoeuvred himself into a logical dead end, as the military means to achieving a limited political purpose cannot be calibrated *by one side alone* to be limited. If the other side musters a large army, one's own political ends might be as limited as they come, but it will still require an army of similar dimensions to defeat the other side if a military victory is part of the political objective. Honig uses the example of the Gulf War, when the American military clearly aimed at the total defeat of Iraq's armed forces while the political objective was limited to the restoration of the status quo, i.e., the liberation of Kuwait from Iraqi occupation. Thus, to the frustration of the military, the UN's Security

Council stopped the coalition's military effort short of the comprehensive annihilation of Iraq's armed forces and, indeed, of the seizure of the enemy's capital, Baghdad. Saddam Hussein remained in power and a good proportion of his armed forces survived the coalition attack on Iraq. The UN achieved its political purpose if we apply the Clausewitzian realist definition of victory as imposing one's will on the enemy: the UN forced Iraq to renege on its will to occupy Kuwait. The Clausewitzian idealist definition of victory did not apply, however: Iraq was not made *defenceless*.

Clausewitz's alleged failure to confront limited war can only be upheld in part, in view of Clausewitz's lectures on 'small war' and his treatment of *guerrilla* warfare which we discussed in Chapter 6. It is a fact, however, that Clausewitz did not pull his different strands of thinking together into a comprehensive synthesis.

Following van Creveld, Jan Willem Honig also points to the divergence between Western 'military strategies . . . based on a capacity to escalate and destroy' that is at odds with the increasing number of cases in which the West's political stakes are too limited to warrant the employment of overwhelming force.[13] The general rule seems to be, however, that if one side has lower political stakes in a conflict than its opponent and is thus unwilling politically to escalate the conflict beyond certain limits, the opponent, with a greater political stake, is often able to draw out the conflict on a level of low military intensity until the first side to the conflict tires of its involvement and withdraws. That this is possible even with powers that have the military means to escalate to major or even nuclear war has been demonstrated in America's Vietnam experience and in the USSR's experience in Afghanistan. It could be argued that Clausewitz failed to elaborate sufficiently the third most important rule for the conduct of war which he himself gave to the Crown Prince, namely, to win over public opinion. Clausewitz doesn't solve the riddle of whether it is one's own or the enemy's public opinion one has to win, or perhaps even 'world opinion', a subject on which Clausewitz's colleague Rühle von Lilienstern was much more articulate.[14]

Liddell Hart, whom we have already encountered as one of the bitterest critics of Clausewitz, further faulted him for not giving enough consideration to his own narrow definition of strategy ('the use of engagements for the object of the war') – the level that was later called 'grand strategy' in English usage. He also criticised him for not giving more thought to the peace which must follow war.

> Fighting power is but one of the instruments of grand strategy – which should take account of and apply the power of financial pressure, and, not least of ethical pressure, to weaken the opponent's will . . . Furthermore, while the horizon of strategy is bounded by the war,

grand strategy looks beyond the war to the subsequent peace. It should not only combine the various instruments, but so regulate their use as to avoid damage to the future state of peace – for its security and prosperity.[15]

It is perhaps equally unfair to denounce Liddell Hart's attempt to resist the all-pervasive quest for 'absolute war' as appeasement of National Socialism. As Azar Gat has observed, while Clausewitz discovered the political rationale for limited war in 1827,

> the implications and practical strategies for that type of war were left to be worked out by later societies, for which limiting war was again becoming not just a *theoretical* option. For these societies, military 'victory' increasingly appeared, while not disconnected from, more and more difficult to identify with a successful peace and an overall gain – a notion still alien to Clausewitz. A major reformulation of strategic theory was thus required, was first attempted, even before the First World War, by Julian Corbett – another Edwardian liberal – and was then developed during the inter-war period by Liddell Hart.

For Gat, then, Corbett and Liddell Hart promoted the heritage of Clausewitz the Realist's limited-war thinking, in the interest of containing the destruction wrought by war.[16]

With the same quest for limitations of war in mind, the American strategist and military historian Edward Luttwack has argued that in the post-Cold War era it is important to return to a pre-Napoleonic form of war, taking our orientation from the form of war which Clausewitz the Idealist had regarded as obsolete. Luttwack writes that the West would do well to 'emulate the casualty-avoiding methods of eighteenth century warfare and thus conduct armed yet virtually bloodless interventions.'[17]

The latest Chichele Professor of the History of War in Oxford Hew Strachan is convinced that this is already the case: he points to a development in the Western way of war away from absolute, let alone total, war. Since the First World War, and particularly after 1945, it has been the great powers, and later the nuclear powers who, thanks to their technological superiority in other areas also, could aim to avoid major war by fighting very unequal wars against technologically inferior adversaries, resulting in ever more limited casualty figures for them. It was societies unable to throw sufficient high technology into the scales of war who felt obliged to compensate by an ever more comprehensive mobilisation of their own population. Towards the end of the twentieth century, warfare became a remote spectator sport for the US, Britain and France, even though the casualties they could inflict on their adversaries remained high.[18] These developments bring the gap in Clausewitz's writing on such asymmetrical wars sharply into focus.

There are also concrete points in which the development of technology has simply changed some of the basic factors of the early nineteenth century. While Clausewitz thought tactical surprise a good means of undermining the enemy's morale, and strategic surprise 'the most effective principle for victory', he thought it extremely difficult to accomplish in view of the secret preparations needed.[19] In the nuclear age the 'bolt from the blue' strategic nuclear attack of course became the worst nightmare of military planners both in NATO and in the WTO. Clausewitz thought little of intelligence (which is perhaps excusable given the means available in his day and age).[20] With new technology intelligence became increasingly vital in the twentieth century.[21] As we have seen, Clausewitz did acknowledge the impact of evolving technology on warfare but made little allowance for it elsewhere in his reflections.[22]

Eternal War or Eternal Change?

Perhaps the most philosophically profound problem with Clausewitz is his quest for identifying eternal truths about war, while at the same time conceding that all wars differ according to the circumstances of their time. In other words, there is a fundamental methodological problem with the relationship Clausewitz tried to construct between his database of historical wars and the general theory of war which he tried to distil from it. He was not alone with this problem.

Gat sees here a profound debt to Scharnhorst in Clausewitz's approach.[23] Indeed, most Prussian/German strategists would have agreed at the time that 'while the forms (*Formen*) of war may change with time, its spirit (*Geist*), or essence (*Wesen*), remains unchanged.'[24] We also find this in French strategic literature. Antoine de Jomini, in his *Summary of the Art of War* of 1837, famously claimed:

> There exist a small number of fundamental principles of war, and if they are found sometimes modified according to circumstances, they can nevertheless serve in general as a compass to the chief of any army . . . Natural genius will doubtless know how, by happy inspirations, to apply [these] principles as well as the best studied theory could do it . . . [25]
> The fundamental principles upon which rest all good combinations of war have always existed . . . These principles are unchangeable; they are independent of the nature of the arms employed, of times and places . . . For thirty centuries there have lived generals who have been more or less happy in their application.[26]

Strategy particularly may be regulated by fixed laws resembling those of the positive sciences.[27]

Clausewitz himself studied many earlier examples of outstanding generalship. Early on in his professional life, Clausewitz reflected 'on the Campaigns of Gustavus Adolphus of 1630–32'. Hans Rothfels believed that Clausewitz had deliberately chosen to study this early campaign because he opposed an interpretation of the art of war which simply regarded what was most recent as best or even as the only true art of war, from which everything else was a deviation. In this fragment on Gustavus Adolphus we even find the remark that 'the different great wars constitute as many periods in the history of the art [of war]' and should be taken equally seriously.[28]

As we have seen in Chapter 2, Clausewitz was initially guilty himself of selecting too limited a database, in concentrating to the exclusion of everything else on French Revolutionary and Napoleonic warfare in his idealistic first draft of *On War*, and making that 'a universal yardstick' for all war.[29] His quest for a universal theory, however, is clear in Book VIII:

> But there must be something more general or rather something universal in the conduct of war, which is caused by the peculiar conditions of the states [involved] and the armed forces. It is with this that [our] theory will be mainly concerned.
>
> Recent times, in which war attained its absolute strength, saw the most of this generally valid and necessary [characteristic of war]. Yet it is just as unlikely that wars will henceforth have all these great characteristics as that the large barriers which have been opened for them, can be shut again entirely.[30]

We should recall here the long historical survey of war in Book VIII, Chapter 3B, of *On War*, which Clausewitz concludes by explaining that he has not given it

> in order quickly to note down for each period a few principles of the conduct of war, but merely in order to show how each period had its own wars, its own limiting conditions, its own bias. Each would thus also retain its own theory of war, even if one had been everywhere . . . inclined to treat it according to philosophical principles.[31]

Yet even when he widened his scope and in his post-1827 revisions sought to consider wars other than those of the French Revolution and Napoleon, he limited his 'database' to wars which had taken place in modern times. He argued:

The most recent history of war must always be the most natural area from which to choose examples. This is not only because more distant periods are characterised by different circumstances, and thus by a different way of waging war, so that its results are less informative and practical for us.

He also reckoned that with the passing of time, fewer details about individual campaigns and wars were known and could be reconstructed. Thus his own 'database' did not go back much beyond the Thirty Years War. 'The further one goes back in the history of war, the less use it is, because it becomes at once more threadbare and poor. Ancient history must be the least useful and the poorest.'[32]

We thus find time and again in Clausewitz's writing the tension between the Montesquieuean (and later, Rankean) belief in the particularity of every period of war and the quest for some universally true reflections about war. Gat aptly summarises Clausewitz's aspiration: 'Above historical study and crude rules' to identify 'a universal theory which reflects the lasting nature of war, transcends the diversity and transformations of past experience, and is both generally valid and instructive'.[33] As Clausewitz himself wrote in his fragment, *Strategie*, of 1808:

> The example is the real-life case, the formula the abstraction. When by the abstraction nothing which is essential gets lost, as is the case in mathematics, [the abstraction] fully achieves its purpose. But when it must constantly omit the real-life case in order to cling to the dead form . . . it would be in the end a dry skeleton of insipid truths and common places, squeezed into a school curriculum. One has to be surprised to find people who waste their time on such deliberations, when one bears in mind that precisely that which is the most important in war and strategy, namely what is most peculiar and particular to it, and local circumstances, elude these abstractions and scientific systems most.[34]

In changing from the Idealist who claimed to be describing the essence of war as found in French Revolutionary and Napoleonic warfare, to the Realist who admitted that wars could take on all sorts of forms, from very limited to well-nigh absolute, Clausewitz faced a further problem that he identified early on. After 1827, he was indeed sometimes in danger of becoming trivial in his attempts to characterise the 'true spirit of war' in ways that would apply to *all* warfare in *all* periods of history. Back in 1808 he had written:

> The more I think about this part of the art of war, the more I become convinced that its theory can formulate few or no abstract theorems; but

this is not, as is usually supposed, because it would be too difficult to do so, but because they would be trivial. There are so many small factors which in war contribute to determine the action that one would appear like a great pedant and would become disgustingly trivial if one were to include all of them in the abstract theorems. All authors who in recent times have tried to treat this part of theory in an abstract and philosophical way furnish good evidence of this; for they have either become truly trivial, or they have escaped triviality through one-sidedness . . . All that this part of the art of war can aspire to is to reason about the true spirit of war . . .[35]

And he himself commented on his writings of 1816–18: 'I wanted at all costs to avoid every commonplace, everything self-evident that has been stated a hundred times and is generally accepted; for it was my ambition to write a book which will not be forgotten after two or three years.'[36] He was not always entirely successful in avoiding commonplaces, as when writing, 'The first rule would thus be to field the largest possible army. This may sound a platitude but in reality it is not.'[37] One wonders.

How far was Clausewitz able to articulate more meaningful universal theories? At times, he seemed to give up hope. In Book II, Chapter 2 of *Vom Kriege*, he formulated a section heading: 'A positive doctrine is impossible';[38] in Book II Chapter, 5 he wrote: 'All positive results of theoretical examinations, all principles, rules and methods lack general application and absolute truth the more they become positive doctrine.'[39] And elsewhere, 'Nor can the theory of war apply the concept of law to action, since no rule universal enough to deserve the name of law can be applied in view of the constant change and diversity of the phenomenon of war.'[40]

Here we come back to Clausewitz's purpose in writing *On War*, which was to further critical thought and study (*Kritik*) rather than to formulate a generally applicable doctrine.[41] As he wrote in another heading in Book II, Chapter 2: 'Theory should be study and not a doctrine.'[42] Clausewitz thus tried to surmount both the prescriptive type of literature on war (that concerns mostly the 'art' of war) and the theoretical literature, which establishes principles which are either trivial or rules that have too many exceptions to be satisfactory. Clausewitz tried to get to the *essence* of war, its *spirit*, its true *nature*, the 'concept [*Begriff*] of war itself'.[43]

Today we might say that Clausewitz sought to work out a formula containing all the variables that make up war, and his initial formula – violence, genius of the commander, morale of the forces, concentration of forces against the centre of gravity, aim of overthrow of the enemy, etc. – was found by him in 1827 to be deficient as the crucial variable of political aim is missing. This is the variable he introduced in his revisions of 1827–30.

Clausewitz's dithering between the assertion that the true nature of war is eternal and the finding that in reality, every war is different, war is a true chameleon, makes him attractive to both 'International Relations Realists',[44] who believe that strategy never changes (and that the basic patterns of international relations never change), and to those who stress the variability of war and strategy, as they are a function of political aims which in turn they see as a function of different cultures and values, threat perceptions and unique sets of circumstances. Thus on the one hand Colin Gray, who defines himself as a 'Realist', asserts that 'the need to use or threaten force for political objectives, the need to behave strategically, is perennial and universal.' (He for one would be unable to explain why the member states of the EU no longer obey this rule in their behaviour towards each other.) On the other hand, inspired by Clausewitz's Book VIII of *On War*, Bernard Brodie in *War and Politics* devotes a chapter to changing attitudes towards war. While he overestimated the continuity in values and beliefs from antiquity until the French Revolution, he justly highlights the changes that have occurred since then – for example, the importance of concepts such as 'honour' even in the early twentieth century and their deprecation by the middle of that same century – and concomitantly proves the inadequacy of all international relations theories which are posited on the immutability of interests, values and beliefs.[45]

Clausewitz's Enduring Relevance

Mid-Cold War Soviet criticism of Clausewitz included the conviction that he 'had a wrong, idealistic notion of politics, which he saw as the rationality of a personified state. Moreover, he thought of all politics as foreign policy.'[46] His Soviet critics had a point – in the largest part of *On War*, Clausewitz wrote as though there were just one source of policy-making, even if he went beyond the 'black box' notion of the state and identified the components of the classical state of his age (government, people, military).

Strong criticism of Clausewitz also came from the liberal camp. Martin van Creveld had written in the mid-1980s: 'Among the better-known writers on military theory within Western civilisation it is Clausewitz alone whose work appears able to withstand every kind of political, social, economic, and technological change since it was published, and seems to stand a fair chance of remaining forever of more than purely historical interest.'[47] In 1991, however, he wrote: 'Some of the most fundamental problems posed by war in all ages [are] by whom it is fought, what it is all

about, how it is fought, what it is fought for, and why it is fought ...
Contemporary "strategic" thought about every one of these problems is
fundamentally flawed; and, in addition, is rooted in a Clausewitzian world
picture that is either obsolete or wrong' because it is excessively focused on
classical states as warring entities.[48]

Van Creveld pointed out, moreover, that Clausewitz assumes that wars
are always a means to an end and not an end in themselves, while in reality,
'People very often take up one objective or another precisely in order that
they may fight.'[49] This could be the case for a host of different reasons,
from the prestige and power it gives to military leaders, to the economic
impact on the arms industry, arms trade, and so on. Thus UN forces
deployed in Bosnia-Hercegovina and in Kosovo in the mid-to late 1990s
found that the pacification of these two apparently tiny scraps of territory
was far from easy, as there were increasing vested interests among locals
with Mafia-type connections in perpetuation of the state of anarchy.

Making a similar criticism, the British political scientist Chris Brown
thinks that there is a 'fundamental problem with the Clausewitzian account
of war, which is that it may be culturally specific.'

> Nineteenth-century European war was a very formal business, with
> uniformed armies occupying clearly delineated territory, a code of
> conduct which was usually (although not always) observed, a formal
> declaration and a formal end, the peace treaty. The 'decisive battle' was
> a feature of Napoleonic, Clausewitzian and Victorian accounts of war
> ... States fight in a formal way and make peace in a formal way. Hanson
> calls this *The Western Way of War*[50] and traces it back to the wars of the
> classical Greek cities, in which citizen heavy infantry would fight one
> highly stylised battle per campaigning season, with a clear-cut way of
> determining winners and losers based on possession of the battlefield.
> This, he suggests, gives modern Europe its governing idea of what a war
> is like. However ... it is highly untypical of the warfare of most
> civilisations which is much more informal, is not dominated by set-piece
> battles, rarely leads to any kind of decisive moment, much less a peace
> treaty.[51]

Is Clausewitz thus too culturally specific in his idealistic fixation on
decisive battles? The Gulf War of 1990–91 and the tank battles in the later
1990s between Ethiopia and Eritrea showed that 'European style' large-
scale tank battles could still find themselves transposed to other parts of
the world. Did Clausewitz really assume that it was always states with well-
segregated governments, armed forces and general populations who
engaged in war? He would have been the first to recognise that this was not
the case in his own teaching on the *guerrilla* and on the people's uprising.

Clausewitzian concepts, such as 'friction', transcend every mentality and culture.[52] Clausewitz's idea of the 'economy of forces' deserves to be rediscovered and applied, but while this is possible for population-rich states like the USA, India or China, it is difficult for the United Kingdom, with its chronic overstretch. The genius of the commander is still needed to digest ever greater masses of information in technological circumstances in which ever greater political micro-management of warfare is possible. The centre of gravity is as difficult to identify clearly today as it was for Clausewitz, but he opened the door to the realisation that, ultimately, any contest of wills is a battle for the hearts and minds of the enemy, and a struggle to have public opinion everywhere on one's own side, something which Mao particularly stressed.

Parts of the world-view, particularly of Clausewitz the Idealist and his vision of 'absolute war', have been overtaken by the industrialisation of genocide which the twentieth century witnessed, and the perfection of fire power with nuclear weapons. The question still remains in what circumstances wars tend towards absolute violence, and whether 'limited wars' result inevitably from a mutual hesitation to resort to 'the absolute weapon'. One of the British government's scientific advisers, the mathematician Sir Herman Bondi of King's College, London, remarked that a nuclear power is a power nobody can afford to make desperate.[53] While this seems a reasonable proposal, we can only hope that this is clear to decision-makers everywhere.

The limited wars of the eighteenth century were no longer possible in Europe by the end of the nineteenth, while they were still typical, as Corbett stressed, in competition for far-flung colonial possessions. The twentieth century saw limited wars by proxy between nuclear powers concerned to keep violence below the nuclear threshold. But as missile and nuclear technology spreads in the early twenty-first century, it is an article of faith rather than of conviction to say that the possession of all weapons of mass destruction will make their owners more cautious in their military engagements. Several states in the Middle East that either have their own nuclear programmes or are aspiring to them do not stand out as particularly restrained military actors. Meanwhile we find that, unlike in the times of Guibert, Clausewitz and even Corbett, when states could count on the relative readiness of their professional soldiers to engage without demur in limited wars far from their homeland, with the spread of public involvement in politics, states have an ever greater need to obtain the backing of their own public for such operations.

While we find Clausewitz's postulates have limited applicability to decision-making in foreign policy and war, it remains sensible for collective decision-makers to take heed of his advice to think of military options as subordinate to greater political aims, and to take into account the ultimate danger of escalation when taking the first step in the direction of military action. This is good practical advice, even if it is not a ubiquitously applicable analytical key to the understanding of other actors. Clausewitz's own realisation that all wars are functions of the culture, age and mentality in which they were conceived is a warning not to put excessive emphasis on an adversary's putative political masterplan when his culture is simply one of violence and bellicosity. These insights from Book VIII of *On War*, which encourage us to study the particular culture and mind-frame we have ourselves or encounter in an adversary, deserve to be the guiding question of anthropological research into war and society in the twenty-first century. Gaston Bouthoul (1896–1980), the French inventor of the discipline of 'Polemology', said, 'If you want peace, know war.' Clausewitz might have paraphrased him, saying 'If you want security for yourself, know with what aims and in which manner others might go to war.' For the study of war, more than any prescription of how to wage it, was the ultimate purpose of Clausewitz's writing. With the formulation of theory, he aimed to develop 'aids to judgement' as the basis of study, not of normative behaviour.[54]

Unlike writers such as Henry Lloyd or Berenhorst, or indeed Jomini, it was thus not Clausewitz's aim to produce formulas for the successful conduct of battles. Clausewitz's main aim was to understand the nature of war, just as Newton sought to understand the nature of mechanics, not to deduce prescriptions for the construction of trains. It is true, however, that the latter is greatly facilitated by the understanding of the former, just as (to use his metaphor) the guiding stars identified by Clausewitz help the military or civilian decision-maker find his bearings, even if Clausewitz does not tell him whether to turn right or left at any given crossroads.

Unlike Newton, and notwithstanding the list of basic principles which we have cited in Chapter 1, Clausewitz was not able to produce a simple list of axioms with universal applicability. While the formula for the essence of war itself remains incomplete, Clausewitz rightly identified a number of functions, dependent on variables, which are somehow *part* of this formula. To recapitulate, they include the centrality of violence and the difficulties of checking its escalation; the importance of political purpose for the formulation of military aims (ignoring which produces disastrous consequences, as we saw in the First World War); the need for good relations between the supreme civilian authority and the military, and for popular support of any given cause; the talent of the supreme decision-makers (and particularly of the military commander-in-chief) for an intuitive understanding of highly complex situations; the morale of armed forces and populations and of their determination (will) to overcome obstacles; the need to concentrate forces on a centre of gravity; and, finally, the ubiquity of friction. So, mathematically put, the nature of any individual war is a function of its tendency towards escalation, of the rapport between government and military leadership, of popular support for the cause of the war, etc. All these factors, acting differently upon both sides in a conflict, determine the nature of the conflict, its limitations and its violence.

Since Clausewitz's times, a number of other variables have been identified: the military-industrial potential of a fighting entity, the means – arms, raw materials, manpower, a propaganda machinery, etc. – that are available to it, the power of its weapons systems and the traditions which have become established with regard to their use, to name just a few. Clausewitz had thus not detected, let alone explored, all the variables in the system. But – and this is Clausewitz's great achievement, as he himself contested – by teaching us to look for functions and variables determining the nature of all warfare and every particular war, Clausewitz has taught us *how* to think about war, which in turn helps us identify some of the other variables. And with regard to this self-defined purpose, Clausewitz has been more successful than just about any strategist before or after him.

Thus other strategic thinkers, from Corbett to Mao, from Lenin to Brodie, from Moltke to Summers, could develop further the approach which Clausewitz had formulated, and could introduce structure and clarity into their own arguments. It is this approach rather than individual elements of Clausewitz's findings which has proved remarkably resilient to change in society, technology, political systems, and thus warfare, since his times. Some of his reflections on individual variables and their importance as factors determining the nature of any real war are precious, some are rather more dated, and none, taken on their own, are totally original. But fitted together into the intellectual construction of war as a function of these variable factors, Clausewitz's contribution to our thinking on war is outstanding, and has rightly been a model copied by other strategic thinkers since.

Select Bibliography
of Works in English

(For references to works in other languages, please refer to the endnotes)

Aron, Raymond: *Clausewitz, Philosopher of War*, transl. Christine Booker and Norman Stone (London: Routledge & Kegan Paul, 1976)

Bassford, Christopher: 'John Keegan and the Grand Tradition of Thrashing Clausewitz', *War in History*, Vol. 1, No. 3 (1994)

Bassford, Christopher: 'Landmarks in Defense Literature: *On War* by Carl von Clausewitz', *Defense Analysis*, Vol. 12, No. 2 (1996)

Bassford, Christopher: *Clausewitz in English: The Reception of Clausewitz in Britain and America, 1815–1945* (Oxford: Oxford University Press, 1994)

Bond, Brian: *Liddell Hart: A Study of his Military Thought* (London: Cassell, 1977)

Brodie, Bernard: *Strategy in the Missile Age* (2nd ed., Princeton, N.J.: Princeton University Press, 1965)

Brodie, Bernard: *War and Politics* (New York: Macmillan, 1973)

Caemmerer, [Rudolf] von: *The Development of Strategical Science during the 19th Century*, transl. Karl von Donat (London: Hugh Rees, 1905)

Challener, Richard D.: *The French Concept of the Nation in Arms* (New York: Columbia University Press, 1975)

Cimbala, Stephen J.: *Clausewitz and Chaos: Friction in War and Military Policy* (New York: Praeger, 2000)

Cimbala, Stephen J.: *Clausewitz and Escalation: Classical Perspectives on Nuclear Strategy* (London: Frank Cass, 1991)

Corbett, Julian S.: *Some Principles of Maritime Strategy* (London: Conway, 1972)

Deist, Wilhelm *et al.* (Militärgeschichtliches Forschungsamt, Freiburg): *Germany and the Second World War*, Vol. 1: *The Build-up of German Aggression* (Oxford: Clarendon Press, 1990)

Dexter, Byron: 'Clausewitz and Soviet Strategy', *Foreign Affairs*, Vol. 29, No. 1 (Oct. 1950)

Earle, Edward Mead (ed.): *Makers of Modern Strategy from Machiavelli to Hitler* (Princeton, N. J.: Princeton University Press, 1943)

Foch, Ferdinand: *Principles of War*, transl. Hilaire Belloc (London: Chapman & Hall, 1918)

Frank, Willard and Philip Gilette (eds): *Soviet Military Doctrine from Lenin to Gorbachov, 1915–1991* (Westport, Conn.: Greenwood, 1992)

Garthoff, Raymond L.: *Soviet Military Doctrine* (Glencoe, Ill.: Free Press, 1953)

Gat, Azar: *The Origins of Military Thought from the Enlightenment to Clausewitz* (Oxford: Clarendon Press, 1989)

Gat, Azar: *The Development of Military Thought: The Nineteenth Century* (Oxford: Clarendon Press, 1992)

Glantz, David M.: *The Military Strategy of the Soviet Union – A History* (London: Frank Cass, 1992)

Gray, Colin: *Modern Strategy* (Oxford: Oxford University Press, 1999)

Handel, Michael I. (ed.): *Clausewitz and Modern Strategy* (London: Frank Cass, 1986)

Handel, Michael I.: *Masters of War: Classical Strategic Thought* (2nd edn, London: Frank Cass, 1996)

Honig, Jan Willem: 'Interpreting Clausewitz', *Security Studies*, Vol. 3, No. 3 (Spring 1994), pp. 571–580

Howard, Michael and Peter Paret (eds): *Carl von Clausewitz: On War* (Princeton, N.J.: Princeton University Press, 1976)

Howard, Michael: *Clausewitz* (Oxford: Oxford University Press, 1983)

Jomini, Antoine de: *Summary of the Art of War* transl. O.F. Winship and E.E. McLean (New York: Putnam & Co, 1854)

Kahn, Herman: *On Thermonuclear War* (Princeton, N.J.: Princeton University Press, 1970)

Liddell Hart, Basil Henry: *The Ghost of Napoleon* (London: Faber & Faber, n.d. [1934?])

Luvaas, Jay: *The Education of an Army: British Military Thought, 1815–1940* (London: Cassell, 1975)

Mitchell, Alan: *Victors and Vanquished: The German Influence in Army and Church in France after 1870* (Chapel Hill: University of North Carolina Press, 1984)

Murray, Williamson, MacGregor Knox and Alvin Bernstein, *The Making of Strategy: Rulers, States, and War* (New York: Cambridge University Press, 1994)

Osgood, Robert Endicott: *Limited War: The Challenge to American Strategy* (Chicago: University of Chicago Press, 1957)

Osgood, Robert Endicott: *Limited War Revisited* (Boulder, Colorado: Westview Press, 1979)

Otte, T.G.: 'Educating Bellona: Carl von Clausewitz and Military Education', in Keith Nelson and Greg Kennedy (eds): *Military Education: Past, Present, Future* (New York: Praeger, 2002)

Paret, Peter: *Clausewitz and the State: The Man, His Theories, and His Times* (rev. ed., Princeton: Princeton University Press, 1985)

Paret, Peter (ed.): *Makers of Modern Strategy from Machiavelli to the Nuclear Age* (Princeton, N. J.: Princeton University Press, 1986)

Reid, Brian Holden: *J.F.C. Fuller: Military Thinker* (London: Macmillan, 1987)

Ritter, Gerhard: *The Schlieffen Plan: Critique of a Myth*, transl. Andrew and Eva Wilson (London: Wolff, 1978)

Roger Chickering: 'Total War: The Use and Abuse of a Concept', in Manfred Boemeke, Roger Chickering and Stig Förster (eds): *Anticipating Total War: The German and American Experiences, 1871–1914* (Cambridge: Cambridge University Press, 1999)

Snyder, Jack: *The Ideology of the Offensive: Military Decision-Making and the Disasters of 1914* (Ithaca: Cornell University Press, 1984)

Strachan, Hew: 'On Total War and Modern War', *International History Review*, Vol. 22, No. 2 (June 2000)

Summers Jr, Colonel Harry G.: *On Strategy: A Critical Analysis of the Vietnam War* (Novato, Ca.: Presidio, 1982)

Wallach, Jehuda L.: *The Dogma of the Battle of Annihilation: The Theories of Clausewitz and Schlieffen and their Effect on the German Conduct of Two World Wars* (Westport, Conn.: Greenwood, 1988)

Zedong, Mao: *Selected Military Writings* (Beijing: Foreign Languages Press, 1977)

Notes

Foreword

1. Eberhard Koenig (ed.), *Sophie Gräfin Schwerin: Ein Lebensbild aus ihren zueigenen hinterlassenen Papieren, zusammengestellt von ihrer jüngsten Schwester, Amalie von Romberg* (Leipzig, 1911 ff), vol. 3.

1. The Story of the Man and the Book

1. He is thus *not* called Carl Maria, as even some of his greatest fans, such as Colin Gray, believe, which would have made him a Bavarian or Austrian Catholic and not a Prussian Protestant. See Colin Gray: *Modern Strategy* (Oxford: Oxford University Press, 1999), p.26. This is particularly piquant in view of Gray's emphasis on the importance of culture for strategy, see pp.129–51.
2. 'News about Prussia in its great catastrophe' (1823–4), quoted in Eberhard Kessel (ed.): *Carl von Clausewitz – Strategie aus dem Jahr 1804 mit Zusätzen von 1808 und 1809* (Hamburg: Hanseatische Verlagsanstalt, 1943), p.9.
3. Clausewitz: 'Historisch-politische Auszüge und Betrachtungen' (1803), in Hans Rothfels (ed.), *Carl von Clausewitz: Politik und Krieg, Eine Ideengeschichtliche Studie* (Berlin: Dümmler, 1920), p.197.
4. Clausewitz in 1808, in *ibid.*, p.211.
5. Hans Rothfels (ed.), *Carl von Clausewitz: Politische Schriften und Briefe* (Munich: Drei Masken Verlag, 1922), pp.49–51.
6. Carl von Clausewitz: 'Über einen Krieg mit Frankreich', in Karl Schwartz (ed.): *Leben des Generals . . . Carl von Clausewitz und der Frau Marie von Clausewitz geborene Gräfin von Brühl mit Briefen, Aufsätzen, Tagebüchern und anderen Schriftstücken* (Berlin: Dümmler, 1878), Vol. 2, pp.418–39.
7. Clausewitz to Gneisenau, 21 Oct. 1830, quoted in G. H. Pertz and Hans Delbrück: *Das Leben des Feldmarschalls Grafen Neithardt von Gneisenau* (Berlin: 1864ff.), Vol. V, p.609, transl. in Peter Paret, *Clausewitz and the State* (Oxford: Clarendon Press, 1976), p.399.

8. Caroline von Rochow, an acquaintance of the Brühls, quoted and transl. in Paret: *Clausewitz and the State*, p.104.

9. Karl Linnebach (ed.): *Karl und Marie von Clausewitz: Ein Lebensbild in Briefen und Tagebuchblättern* (Berlin: Martin Warneck, 1917), pp.76–149.

10. 'Bekenntnisdenkschrift' (1812), in Rothfels (ed.), *Politische Schriften und Briefe*, p.85.

11. Eugène Carrias: *La pensée militaire allemande* (Paris: Presses Universitaires de France, 1948), p.186.

12. Carl von Clausewitz: *Hinterlassene Werke über Krieg und Kriegführung*, Parts I–III (Berlin: Ferdinand Dümmler Verlag, 1832–34).

13. 'Nachricht vom 10. Juli 1827', *Vom Kriege*, p.180f.

14. All quoted in Werner Hahlweg: *Carl von Clausewitz: Soldat, Politiker, Denker* (Göttingen: Musterschmidt Verlag, 1957), p.29f.

15. All quoted in Friedrich Doepner: 'Clausewitz als Soldat', *Europäische Wehrkunde*, Vol. 29, No. 7 (July 1980), pp.351, 354.

16. *Ibid., passim.*

17. Peter Paret: *Clausewitz and the State* (Oxford: Clarendon Press, 1976), p.151.

18. Rothfels (ed.): *Politik und Krieg*, p.29f.

19. *Vom Kriege*, Preface, p.175.

20. Hermann Cohen: *Von Kants Einfluss auf die deutsche Kultur, Marburger Universitätsrede* (Berlin: 1883), p.31f, quoted in Rothfels (ed.): *Politik und Krieg*, pp.23–5.

21. Walther Malmsten Schering: *Wehrphilosophie* (Leipzig: Johann Ambrosius Barth, 1939), pp.89–93 (CvC and Kant) and 93–101 (CvC and Hegel).

22. Lenin in *The Collapse of the 2nd Internationale*, quoted in Major General Professor Rasin: 'Die Bedeutung von Clausewitz für die Entwicklung der Militärwissenschaft', *Militärwesen* 2nd year, No. 3 (May 1958), p.379.

23. Letter to Fichte (1809), in Werner Hahlweg (ed.): *Carl von Clausewitz: Verstreute kleine Schriften* (Osnabrück: Biblio Verlag, 1979), pp.159–66.

24. In Rothfels (ed.): *Politische Schriften und Briefe*, p.64.

25. Niccolò Machiavelli: *The Prince* (transl. George Bull, Harmondsworth: Penguin, 1961), pp.119–24.

26. 'Bekenntnisdenkschrift' (1812), in Rothfels (ed.): *Politische Schriften und Briefe*, p.85.

27. Quoted in K. Schwartz: *Leben des Generals Carl von Clausewitz und der Frau Marie von Clausewitz geb. Gräfin von Brühl* (Berlin: 1878), Vol. 2, pp.290–91, transl. in Peter Paret: *Clausewitz and the State* (Oxford: Clarendon Press, 1976), p.294.

28. Hermann Stodte: *Die Wegbegleiter des National-Sozialismus* (Lübeck: H. G. Rahtgens, 1936), p.52.

29. Paret: *Clausewitz and the State*, p.439.

30. Carl von Clausewitz: 'Umtriebe' (dated between 1819 and 1823), in Rothfels (ed.): *Politische Schriften und Briefe*, p.171.

31. Bülow: *Geist des neuern Kriegssystems*, transl. as *The Spirit of the Modern*

System of War (London: 1806), p.184, quoted in Azar Gat: *The Origins of Military Thought* (Oxford: Clarendon Press, 1989), p.85.

32. Clausewitz: 'Bemerkungen über die reine und angewandte Strategie des Herrn von Bülow . . .', in Hahlweg (ed.): *Verstreute Kleinere Schriften*, pp.65–88.
33. Gat: *The Origins*, pp.95–105.
34. Scharnhorst: 'Entwicklung der allgemeinen Ursachen des Glücks der Franzosen in dem Revolutionskriege', *Neues Militärisches Journal*, Vol. VIII (1797); reprinted in Colmar von der Goltz (ed.): *Militärische Schriften von Scharnhorst* (Berlin: F. Schneider & Co., 1881), pp.199–203.
35. Scharnhorst: 'Nutzen der militärischen Geschichte, Ursach ihres Mangels' (1806), printed in Ulrich von Gersdorff (ed.): *Scharnhorst – Ausgewählte Schriften* (Osnabrück: 1983), pp.199–207; Gat: *The Origins*, pp.165f.
36. Gat: *The Origins*, p.185, n.51.
37. See, e.g., his rejection of 'surprise' as key to success, Book III, Ch. 9. He did regard it as a key to success earlier, see Clausewitz, 'Übersicht des Sr. königl. Hoheit . . . ', *Vom Kriege*, p.1070.
38. *Vom Kriege*, II, 2, p.279.
39. *Vom Kriege*, II, 2, p.290f.
40. Jomini: *Summary of the Art of War*, transl. by O. F. Winship and E. E. McLean (New York: Putnam & Co, 1854), p.14f.
41. On the dating of the note see Gat: *The Origins*, pp.255–63.
42. 'Nachricht', printed in Hahlweg (ed.): *Vom Kriege*, pp.182f.
43. Vom Kriege, 'Nachricht' of 1828, p.180, and II, 2, p.291.
44. Quoted in Ulrich Marwedel: *Carl von Clausewitz: Persönlichkeit und Wirkungsgeschichte seines Werkes bis 1918* (Boppard am Rhein: Harald Boldt Verlag, 1978), pp.108–12.
45. Jomini: *Summary of the Art of War*, transl. Winship/McLean, p.15.
46. Eberhard Kessel: *Moltke* (Stuttgart: K. F. Koehler, 1957), p.108.
47. Jähns, quoted in Marwedel: *Carl von Clausewitz*, p. 119.
48. Karl Marx, Friedrich Engels: *Briefwechsel*, Vol. II, 1854–60 (East Berlin: Dietz Verlag, 1949), p.336.
49. Marx/Engels: *Werke* (1957–68), Vol. 11, p.577, and Vol. 21, p.350, quoted in Marwedel: *Carl von Clausewitz*, p.180.
50. Marx, Engels: *Briefwechsel*, Vol. II, 1854–60, p.339, English in Marx's original.
51. Quoted in Azar Gat: *Military Thought: The Nineteenth Century* (Oxford: Clarendon Press, 1992), p.230.
52. Friedrich Engels: 'The European War', *New York Daily Tribune* (17 Feb. 1855), quoted in Panajotis Kondylis: *Theorie des Krieges: Clausewitz – Marx – Engels – Lenin* (Stuttgart: Klett-Cotta, 1988), p.150.
53. Friedrich Engels in the *Anti-Dühring*, quoted in Panajotis Kondylis: *Theorie des Krieges: Clausewitz – Marx – Engels – Lenin* (Stuttgart: Klett-Cotta, 1988), p.173.

54. Rocquancourt: *Cours complet* (1840), quoted in Marwedel: *Carl von Clausewitz*, p.232.
55. Gat: *The Origins*, pp.128–130.
56. De la Barre-Duparcq: *Commentaires*, quoted in Marwedel: *Carl von Clausewitz*, p.235.
57. Henry Contamine: *La Revanche 1871–1914* (Paris: Berger-Levrault, 1957), pp.25–37.
58. Jules Lewis Lewal: *Stratégie de Combat*, Vol. I (Paris: Librairie militaire de L. Baudouin, 1895), pp.6–8.
59. General Berthaut: *Principes de stratégie* (Paris: Librairie militaire de J. Dumaine, 1881).
60. Douglas Porch: 'Clausewitz and the French, 1871–1914', in Michael Handel (ed.): *Clausewitz and Modern Strategy* (London, Frank Cass, 1986), pp.292f.
61. Gat: *The Nineteenth Century*, p.132.
62. Gat: *The Nineteenth Century*, pp.126–8.
63. Clausewitz: *Théorie de la grande guerre*, transl. de Vatry (1886), Vol. 1, p.x, quoted in Marwedel: *Carl von Clausewitz*, p.237.
64. Quoted in Marwedel: *Carl von Clausewitz*, p.243f.
65. Eugène Carrias: *La Pensée militaire française* (Paris: Presses Universitaires de France, 1960), p.253.
66. Hubert Camon: *Clausewitz* (Paris: R. Chapelot, 1911), p.vi.
67. Admiral Raoul Castex: *Strategic Theories*, transl. Eugenia Kiesling (Annapolis, Md: Naval Institute Press, 1994), pp.xxxviii and 205.
68. Jay Luvaas: *The Education of an Army: British Military Thought, 1815–1940* (London: Cassell, 1965), p.48.
69. John Mitchell: *Thoughts on Tactics and Military Organization* (London: Longmans, Orme, Brown, Green and Longman, 1838), pp.7f.
70. Christopher Bassford: *Clausewitz in English: The Reception of Clausewitz in Britain and America, 1815–1945* (Oxford: Oxford University Press, 1994), p.56.
71. Gat: *The Nineteenth Century*, pp.5–16.
72. G. F. R. Henderson: *The Science of War* (London: Longmans, Green and Co., 1905), p.173.
73. Bassford: *Clausewitz in English*, p.75.
74. J. F. C. Fuller: *The Dragon's Teeth: A Study of War and Peace* (London: Constable, 1932), pp.66f.
75. Quoted in Christopher Bassford: 'John Keegan and the Grand Tradition of Thrashing Clausewitz', *War in History*, Vol. 1, No. 3 (1994), p.331.
76. John Keegan: *A History of Warfare* (New York: Knopf, 1993).
77. Bassford: *Clausewitz in English*, p.74.
78. Russell F. Weigley: *The American Way of War: A History of United States Military Strategy and Policy* (Bloomington: Indiana University Press, 1973), pp.82–4, 95–127.
79. Quoted in Weigley: *The American Way of War*, pp.210f.
80. Bernard Brodie: *Strategy in the Missile Age* (2nd ed., Princeton, New Jersey: Princeton University Press, 1965), pp.34, 36.

81. Bernard Brodie: 'The Continuing Relevance of *On War*' in Michael Howard and Peter Paret (eds): *Carl von Clausewitz: On War* (Princeton, NJ: Princeton University Press, 1976), p.53.
82. Bassford: 'John Keegan', p.320.
83. See below, Chapters 6 and 7.
84. Christian Schmitt: 'Sur le chapitre 18 [du livre III] de la Guerre de C. von Clausewitz', in *Stratégique*, No. 18 (1983), pp.7–23.
85. Alexis Philonenko: 'Clausewitz ou l'oeuvre inachevée: l'esprit de la guerre', *Revue de Métaphysique et de Morale*, Vol. 95, No. 4 (Oct.–Dec. 1990), pp.471–512.
86. Hervé Guineret: *Clausewitz et la guerre* (Paris: Presses Universitaires de France, 1999).
87. Gat: *The Origins*, pp.169f.
88. Tolstoy: *War and Peace*, transl. C. Garnett (London: Heinemann, 1911), p.977.
89. M. I. Dragomirov: *Manuel pour la préparation des troupes au combat*, 3 vols (Paris: Librairie militaire de L. Baudoin, 1886–8); M. I. Dragomirov: *Principes essentiels pour la conduite de la guerre: Clausewitz interprété par le général Dragomiroff* (Paris: L. Baudoin, 1889).
90. Werner Hahlweg: 'Das Clausewitzbild einst und jetzt', in *Vom Kriege*, p.128.
91. Quoted in Raymond Aron: *Clausewitz, Philosopher of War*, transl. by Christine Booker and Norman Stone (London: Routledge & Kegan Paul, 1976), p.401.
92. Gat: *The Nineteenth Century*, p.237.
93. Hahlweg: 'Das Clausewitzbild einst und jetzt', in *Vom Kriege*, pp.128f.
94. Zhang Yuan-Lin: 'Mao Zedong und Carl von Clausewitz: Theorien des Krieges, Beziehung, Darstellung und Vergleich' (MS Ph.D., University of Mannheim, 1995), pp.22–4.
95. *Ibid.*, pp.20–22.
96. Helmut Schmidt: *Menschen und Mächte* (Berlin: Goldmann/Siedler, 1987), p.359.
97. Walther Malmsten Schering: *Wehrphilosophie* (Leipzig: Johann Ambrosius Barth, 1939), pp.22f.
98. P. M. Baldwin: 'Clausewitz in Nazi Germany', *Journal of Contemporary History*, Vol. 16 (1981), pp.10–15.
99. Hermann Stodte: *Die Wegbegleiter des National-Sozialismus* (Lübeck: H. G. Rahtgens, 1936), pp.50–56. See also works quoted by Panajotis Kondylis: *Theorie des Krieges: Clausewitz – Marx – Engels – Lenin* (Stuttgart: Klett-Cotta, 1988), p.10.
100. Quoted in Christopher Bassford: 'Introduction' to Hans W. Gatzke transl.: *Principles of War* (USA: Military Service Publishing Company, 1942), also on Stackpole, www.mnsinc.com/cbassfrd/Stackpol.htm.
101. A. M. Baldwin: 'Clausewitz in Nazi Germany', *Journal of Contemporary History*, vol. 16 (1981), p.10.
102. General Gunther Blumentritt, quoted in Michael I. Handel: *Masters of*

War: Classical Strategic Thought (2nd edn, London: Frank Cass, 1996), p.24.

103. Hans Rothfels: 'Clausewitz', in Edward Mead Earle (ed.): *Makers of Modern Strategy: Military Thought from Machiavelli to Hitler* (Princeton, NJ: Princeton University Press, 1944), p.93.

104. Uwe Hartmann: *Carl von Clausewitz: Erkenntnis, Bildung, Generalstabsausbildung* (Munich: Olzog, 1998), discusses his philosophical and methodological approach at length.

105. For example, Dietmar Schössler (ed.): *Die Entwicklung des Strategie und Operationsbegriffs seit Clausewitz* (Munich: University of the Bundeswehr, 1997).

106. John Gooch: 'Clausewitz Disregarded: Italian Military Thought and Doctrine, 1815–1943', in Michael I. Handel (ed.): *Clausewitz and Modern Strategy* (London: Frank Cass, 1986), p.303.

107. Bassford: 'John Keegan', p.73.

108. Yugo Asano: 'Influences of the Thought of Clausewitz on Japan since the Meiji Restoration', in Clausewitz-Gesellschaft (ed.): *Freiheit oder Krieg? Beiträge zur Strategie-Diskussion der Gegenwart im Spiegel der Theorie von Carl von Clausewitz* (Bonn: Dümmler, 1980), pp.379–94.

109. Quoted in Hahlweg: 'Das Clausewitzbild einst und jetzt' in *Vom Kriege* (19th edn. Bonn: Dümmler, 1991), p.60.

110. Quoted in Marwedel: *Carl von Clausewitz*, p.117.

111. Jähns, quoted in Marwedel: *Carl von Clausewitz*, p.119.

112. Bismarck quoted in Jehuda Wallach: *The Dogma of the Battle of Annihilation* (Westport, Ct: Greenwood Press, 1986), p.198.

113. Quoted in B. H. Liddell Hart: *The Other Side of the Hill* (London: Cassell, 1948), p.203.

114. Marwedel: *Carl von Clausewitz*, p.173.

115. Jay Luvaas: 'Clausewitz, Fuller and Liddell Hart', in Michael I. Handel (ed.): *Clausewitz and Modern Strategy* (London: Frank Cass, 1986), p.199.

116. Liddell Hart: *Defence of the West: Some Riddles of War and Peace* (London: Cassell & Co., 1950), p.293.

117. Luvaas: 'Clausewitz, Fuller and Liddell Hart', p.211.

118. Liddell Hart: *Strategy* (London: Faber & Faber, 1954), p.352.

119. Colin S. Gray: *Modern Strategy* (Oxford: Oxford University Press, 1999), pp.81, 85.

2. Clausewitz the Idealist vs Clausewitz the Realist

1. Carl von Clausewitz: *Vom Kriege* (19th edn., Bonn: Dümmler, 1991), VIII, 3, pp.966–71.

2. Quoted in Eberhard Kessel: *Militärgeschichte und Kriegstheorie in neuerer Zeit: Ausgewählte Aufsätze* (Berlin: Duncker & Humblot, 1987), p.36.

3. Guibert: 'Essay général de tactique', in Comte de Guibert: *Stratégiques* (Paris: L'Herne, 1977), pp.137f.

4. *Vom Kriege*, VIII, 3B, p.963.
5. Johann Gottlieb Fichte: *Die Grundlagen des Naturrechts nach den Prinzipien der Wissenschaftslehre* (the Basis of Natural Law according to the Principles of Scientific Teaching). Quoted in Ernst Hagemann: *Die deutsche Lehre vom Kriege: Von Berenhorst zu Clausewitz* (Berlin: E. S. Mittler & Sohn, 1940), p.103.
6. Fichte: *Machiavell* (ed. H. Schulz, Leipzig: 1918), p.20, quoted in Ernst Hagemann: *Die deutsche Lehre vom Kriege*, p.107, n.1.
7. Clausewitz: 'Preußen in seiner großen Katastrophe', in Hans Rothfels (ed.): *Carl von Clausewitz: Politische Schriften und Briefe* (Munich: Drei Masken Verlag, 1922), pp.202–16 *passim*.
8. 'Übersicht des Sr. königl. Hoheit dem Kronprinzen in den Jahren 1810, 1811 und 1812 vom Verfasser erteilten militärischen Unterrichts', in *Vom Kriege* (19th edn., Bonn: Dümmler, 1991), p.1078.
9. *Vom Kriege*, I, 2, pp.214f.
10. *Ibid.*, I, 2, p.223.
11. *Ibid.*, I, 2, p.229.
12. *Ibid.*, IV, 3, pp.422f.
13. *Ibid.*, IV, 4, pp.427ff.
14. *Ibid.*, IV, 4, p.433.
15. *Ibid.*, IV, 9–11.
16. *Ibid.*, IV, 11, pp.467–71.
17. *Ibid.*, IV, 28, p.814.
18. *Ibid.*, IV, 27, p.808.
19. *Ibid.*, VII, 3, p.875.
20. *Ibid.*, VII, 6, p.881.
21. *Ibid.*, VIII, 9, p.1033.
22. 'Übersicht des sr. königl. Hoheit . . . ', in Halweg (ed.), *Vom Kriege*, p.1070.
23. Aron: *Clausewitz, Philosopher of War*, transl. Booker and Stone (London: Routledge & Kegan Paul, 1976), p.59.
24. Eberhard Kessel: 'Zur Genesis der modernen Kriegslehre: Die Entstehungsgeschichte von Clausewitz's Buch "Vom Kriege"', *Wehrwissenschaftliche Rundschau*, 3rd year, No. 9 (1953), pp.405–23 (this includes the notes written by Clausewitz explaining to a friend how he intended to re-write *On War* from 1827 onwards); Delbrück: 'Ermattungsstrategie' in *Zeitschrift für Preussische Geschichte und Landeskunde*, Vols 11 & 12 (1881), pp.555, 568, 572.
25. *Vom Kriege*, VI, 28, p.813.
26. R[ühle] von L[ilienstern]: *Handbuch für den Offizier zur Belehrung im Frieden und zum Gebrauch im Felde*, Vol. 1 (Berlin: G. Reimer, 1817).
27. R. Von L.: *Handbuch für den Offizier* . . . , Vol. 2, p.8, and see below, Chapter 3.
28. Quoted in Werner Hahlweg: 'Das Clausewitzbild einst und jetzt', in Hahlweg (ed.): *Vom Kriege*, p.87.
29. 'One' here should be taken to mean Clausewitz himself.
30. Clausewitz: 'Gedanken zur Abwehr' (22 Dec. 1827), in Werner Hahlweg

(ed.): *Carl von Clausewitz: Verstreute kleine Schriften* (Osnabrück: Biblio Verlag, 1979), pp.497–9; see also *Vom Kriege*, VIII, 4, p.975.

31. Clausewitz's letter of 21 Nov. 1829 to Gröben, printed in Eberhard Kessel: 'Zur Genesis der modernen Kriegslehre: Die Entstehungsgeschichte von Clausewitz's Buch "Vom Kriege"', *Wehrwissenschaftliche Rundschau*, 3rd year, No. 9 (1953), pp.422f.
32. E.g., *Vom Kriege*, VIII, 1 (Introduction).
33. *Ibid.*, VII, 3, p.875; see also VIII, 7, p.999.
34. *Ibid.*, VIII, 2, p.953.
35. Clausewitz: 'Gedanken zur Abwehr' (22 Dec. 1827), in Werner Hahlweg (ed.): *Carl von Clausewitz: Verstreute Kleine Schriften* (Osnabrück: Biblio Verlag, 1979), pp.495f.
36. Clausewitz: 'Strategie aus dem Jahr 1804', in Werner Hahlweg (ed.): *Carl von Clausewitz: Verstreute Kleine Schriften* (Osnabrück: Biblio Verlag, 1979), p.20.
37. 'Nachricht', in Halweg (ed.), *Vom Kriege*, p.179.
38. *Vom Kriege*, VIII, 2, p.955.
39. *Ibid.*, I, 1, 28, p.212.
40. *Ibid.*, VIII, 6B, p.990.
41. *Ibid.*, VIII, 6B, pp.990ff.
42. *Die Politik* – 'policy' – here could be translated as political decision-makers, those who formulate policy.
43. *Ibid.*, VIII, 6B, p.998.
44. *Ibid.*, I, 1, 27, p.212.
45. *Ibid.*, I, 1, 6, p.196.
46. *Ibid.*, I, 1, 11, pp.200f.
47. *Ibid.*, I, 1, 25, p.211.
48. *Ibid.*, I, 1, 11, pp.200f.
49. *Ibid.*, I, 1, 23 and 24, pp.209f.
50. *Ibid.*, I, 2, pp.216f.
51. *Ibid.*, I, 2, pp.218f.
52. *Ibid.*, I, 2, p.221.
53. 'Strategie aus dem Jahre 1804', in Werner Hahlweg (ed.): *Carl von Clausewitz: Verstreute kleine Schriften* (Osnabrück: Biblio Verlag, 1979), p.35, and *Vom Kriege*, I, 2, pp.223–6.
54. *Vom Kriege*, VIII, 3B, p.962. It is thus incorrect of John Keegan to claim that Clausewitz ignored 'the importance of cultural factors in human affairs'; see John Keegan: *A History of Warfare* (London: Hutchinson, 1993), pp.46f.
55. *Vom Kriege*, VIII, 3B, pp.973f.
56. *Ibid.*, p.953.
57. *Ibid.*, p.973.
58. *Ibid.*, VIII, 2, pp.953ff.
59. *Ibid.*, VIII, 3B, pp.972f.
60. *Ibid.*, VIII, 2, pp.953f.

61. Eberhard Kessel: 'Zur Genesis der modernen Kriegslehre: Die Entstehungsgeschichte von Clausewitz's Buch "Vom Kriege"', *Wehrwissenschaftliche Rundschau*, 3rd year, No. 9 (1953), pp.405–23.
62. *Vom Kriege*, I, 2, p.215.
63. Liddell Hart: *The Ghost of Napoleon* (London: Faber and Faber, n.d., but probably 1934), pp.121, 124–6.

3. Politics, the Trinity and Civil-Military Relations

1. Napoleon: *Maximes* (transl. anon., London: Arthur L. Humphreys, 1906), p.159.
2. Quoted in Ernst Hagemann: *Die deutsche Lehre vom Kriege: Von Berenhorst zu Clausewitz* (Berlin: E. S. Mittler & Sohn, 1940), p.45.
3. Eberhard Kessel: *Moltke* (Stuttgart: K. F. Koehler, 1957), p.35.
4. R(ühle) von L(ilienstern): *Handbuch für den Offizier zur Belehrung im Frieden und zum Gebrauch im Felde*, Vol. II (Berlin: G. Reimer, 1818), pp.8, 13.
5. Karl Marx, Friedrich Engels: *Briefwechsel*, Vol. II, 1854–60 (East Berlin: Dietz Verlag, 1949), p.336.
6. W. I. Lenin: *Clausewitz's Werk 'Vom Kriege', Auszügen und Randglossen* (East Berlin: Verlag des Ministeriums für Nationale Verteidigung, 1957), pp.23–33 *passim*.
7. Quoted in Werner Hahlweg: 'Lenin und Clausewitz', in Günter Dill (ed.): *Clausewitz in Perspektive* (Frankfurt/Main: Ullstein, 1980), pp.636f.
8. Quoted in Clemente Ancona: 'Der Einfluß von Clausewitz' *Vom Kriege* auf das marxistische Denken von Marx bis Lenin', in Günter Dill (ed.): *Clausewitz in Perspektive* (Frankfurt/Main: Leicstein, 1980), pp.582f.
9. A. A. Svechin: 'Evolutsiya strategichesckykh teorii', in B. Garev (ed.): *Voina i voennoe iskusstvo v svete istoricheskogo materializma* (Moscow: Gosizdat, 1927).
10. Quoted in Zhang Yuan-Lin: 'Mao Zedong und Carl von Clausewitz: Theorien des Krieges, Beziehung, Darstellung und Vergleich' (MS Ph.D., University of Mannheim, 1995), pp.135f.
11. *Ibid.*, p.28.
12. Dan Diner: 'Anerkennung und Nichtanerkennung: Über den Begriff des Politischen in der gehegten und antagonistischen Gewaltanwendung bei Clausewitz und Carl Schmitt', in Dill (ed.): *Clausewitz in Perspektive*, pp.447–72.
13. Michel Foucault: *Il faut défendre la société* (Paris: Gallimard-Seuil, 1997), p.44.
14. Emmanuel Terray: *Clausewitz* (Paris: Fayard, 1999), pp.125f.
15. *Vom Kriege*, VIII, 3B, pp.963–5.
16. *Vom Kriege*, VI, 6, 5, pp.640f.
17. Ulrich Scheuner: 'Krieg als Mittel der Politik im Lichte des Völkerrechts', in Clausewitz-Gesellschaft (ed.): *Freiheit ohne Krieg? Beiträge zur*

Strategie-Diskussion der Gegenwart im Spiegel der Theorie von Carl von Clausewitz (Bonn: Dümmler, 1980), pp.159–81.

18. *Vom Kriege*, II, 3, p.303.
19. Peter Paret: 'Die politischen Ansichten von Clausewitz', in Clausewitz-Gesellschaft (ed.): *Freiheit ohne Krieg?*, p.346.
20. On this see Jehuda Wallach: *The Dogma of the Battle of Annihilation* (Westport, Ct: Greenwood Press, 1986), p.51.
21. *Vom Kriege*, VIII, 3B.
22. *Ibid.*, IV, 11, p.470, quoted in Chapter 2.
23. *Ibid.*, I, 1, 3, pp.193f.
24. *Ibid.*, I, 3, p.232.
25. *Ibid.*, III, 6, p.370.
26. *Ibid.*, VI, 6, pp.637f.
27. Text of Clausewitz's essay on the *Landwehr* in Karl Schwartz (ed.): *Leben des Generals Carl von Clausewitz und der Frau Marie von Clausewitz geborene Gräfin von Brühl mit Briefen, Aufsätzen, Tagebüchern und anderen Schriftstücken* (Berlin: Ferdinand Dümmler, 1878), Vol. 2, pp.288ff.
28. Quoted in Hans Rothfels: *Carl von Clausewitz: Politik und Krieg. Eine Ideengeschichtliche Studie* (Berlin: Dümmler, 1920), p.142.
29. Napoleon: *Maximes* (transl. anon., London: Arthur L. Humphreys, 1906), p.159.
30. Christopher J. Duffy: 'The Civilian in Eighteenth-Century Combat', in Erwin A. Schmidl (ed.): *Freund oder Feind? Kombattanten, Nichtkombattanten und Zivilisten in Krieg und Bürgerkrieg seit dem 18. Jahrhundert* (Frankfurt/Main: Peter Lang, 1995), pp.11–29.
31. *Vom Kriege*, VIII, 6B, p.997.
32. *Ibid.*, I, 1, 28, p.213.
33. Aron: *Clausewitz, Philosopher of War* transl. by Booker and Stone (London: Routledge & Kegan Paul, 1976), pp.85f., 93.
34. Edward J. Villacres and Christopher Bassford: 'Reclaiming the Clausewitzian Trinity', *Parameters* (Autumn 1995), also http//www.clausewitz. com/CWZHOME/Trinity/TRINITY.htm; Colin S. Gray: *Modern Strategy* (Oxford: Oxford University Press, 1999), p.28.
35. Quoted in Zhang Yuan-Lin: 'Mao Zedong und Carl von Clausewitz: Theorien des Krieges, Beziehung, Darstellung und Vergleich' (MS Ph.D., University of Mannheim, 1995), pp.138f.
36. *Vom Kriege*, VIII, 6B, p.997.
37. Colonel Harry G. Summers Jr.: *On Strategy: A Critical Analysis of the Vietnam War* (Novato, Ca.: Presidio, 1982).
38. Martin van Creveld: *The Transformation of War* (New York: Free Press, 1991); John Keegan: *A History of Warfare* (New York: Knopf, 1993).
39. Van Creveld: *The Transformation of War*, Ch. 7.
40. *Ibid.*, Ch. 5, 6.
41. *Ibid.*, Ch. 7, and Epilogue.
42. *Ibid.*, Epilogue.
43. Villacres and Bassford: 'Reclaiming the Clausewitzian Trinity'. http:// www.clausewitz.com/CWZHOME/Trinity/TRINITY.htm.

44. *Vom Kriege*, VIII, 6B, pp. 994–6.
45. Carl von Clausewitz: *Vom Kriege* (2nd edn., Berlin: Dümmler, 1853), Vol. 3, p.126.
46. Clausewitz: 'Gedanken zur Abwehr' (22 Dec. 1827), in Werner Hahlweg (ed.): *Carl von Clausewitz: Verstreute Kleine Schriften* (Osnabrück: Biblio Verlag, 1979), p.499.
47. Clausewitz's memorandum on the 'German Military Constitution' of 1815, quoted in Werner Hahlweg: 'Das Clausewitzbild einst und jetzt', in *Vom Kriege*, p.70.
48. *Vom Kriege*, VIII, 6B, p.993.
49. Hans-Ulrich Wehler: 'Der Verfall der deutschen Kriegstheorie', in Ursula von Gersdorff (ed.): *Geschichte und Militärgeschichte* (Frankfurt/Main: Bernard Graefe, 1974), p.289.
50. Raoul Girardet: *La société militaire* (Paris: Librairie Académique Perrin, 1998).
51. Quoted by Werner Hahlweg in *Vom Kriege*, p.67.
52. Bismarck quoted in Wallach: *The Dogma*, p.198.
53. For an excellent analysis, see Stig Förster: 'The Prussian Triangle of Leadership in the Face of a People's War: A Reassessment of the Conflict Between Bismarck and Moltke, 1870–71', in Stig Förster and Jörg Nagler (eds): *On the Road to Total War* (Cambridge, Cambridge University Press, 1993) pp. 115–40.
54. Großer Generalstab (ed.): *Moltkes Militärische Werke*, Vol. I, Pt 1, *Militärische Korrespondenz*, 1864 (Berlin: Mittler, 1864), p.16.
55. Quoted in Gordon Craig: *The Politics of the Prussian Army, 1640–1945* (Oxford: Clarendon Press, 1955, reprinted 1964), p.214.
56. Eberhard Kessel: *Moltke* (Stuttgart: K. F. Koehler, 1957), p.508.
57. Max Horst (ed.): *Moltke: Leben und Werk in Selbstzeugnissen – Briefe, Schriften, Reden* (2nd edn., Bremen: Carl Schünemann Verlag, n.d.), p.361.
58. Craig: *The Politics of the Prussian Army, 1640–1945*, p.195.
59. *Ibid.*, pp.200f., 204.
60. Bismarck: *Gesammelte Werke*, Vol. XV, p.312, quoted in *ibid.*, p.205.
61. Förster: 'The Prussian Triangle', pp.124–34.
62. Craig: *The Politics of the Prussian Army, 1640–1945*, pp.215f.
63. Großer Generalstab (ed.): *Moltkes Militärische Werke*, Vol. IV, *Kriegslehren*, Part 1 (Berlin: Ernst Mittler und Sohn, 1911), p.13.
64. *Ibid.*, pp.13f.
65. Wallach: 'Misperceptions of Clausewitz's *On War*', p.229.
66. Wallach: *The Dogma*, p.39.
67. Wallach: 'Misperceptions of Clausewitz's *On War*', p.216.
68. Ulrich Marwedel: *Carl von Clausewitz: Persönlichkeit und Wirkungsgeschichte seines Werkes bis 1918* (Boppard am Rhein: Harald Boldt Verlag, 1978), pp.153f.
69. Craig: *The Politics of the Prussian Army*, p.218.

70. Quoted and translated in Jehuda Wallach: 'Misperceptions of Clausewitz's *On War* by the German Military', in Michael Handel (ed.): *Clausewitz and Modern Strategy* (London, Frank Cass, 1986), p.228.

71. Colmar, Freiherr von der Goltz: *Jena to Eylau* (London: L. Paul, Trench, Trubner, 1913), pp.75–6.

72. Colmar von der Goltz: *The Nation in Arms* (London: Hodder and Stoughton), p.77.

73. Lieut.-Gen. von Cammerer: *The Development of Strategical Science during the 19th Century*, transl. Karl von Donat (London: Hugh Rees, 1905), pp.85f.

74. Christopher Bassford: *Clausewitz in English: The Reception of Clausewitz in Britain and America, 1815–1945* (Oxford: Oxford University Press, 1994), pp.75f.

75. Général Iung: *Stratégie, Tactique et Politique* (Paris: Charpentier, 1890), pp.260, 287–9, 306.

76. Colin: *Les Transformations de la Guerre* (Paris: Economica, repr. 1989), p.241.

77. Friedrich von Bernhardi: *Vom heutigen Kriege* (Berlin: Ernst Siegfried Mittler & Sohn, 1912), Vol. 2, pp.205, 207f., 434–40.

78. Generalfeldmarschall von Hindenburg: *Aus meinem Leben* (Leipzig: Hirzel, 1920), p.101.

79. Oberst a.D. Bernard Schwertfeger: 'Die politischen und militärischen Verantworktlichkeiten im Verlauf der Offensive 1918', in *Das Werk des Untersuchungsausschusses der Deutschen Verfassungsgebenden Nationalversammlung und des Deutschen Reichstages 1919–1928*, 4th Series, Vol. 2 (1925), p.79, quoted and translated in Wallach: 'Misperceptions of Clausewitz's *On War*', p.231.

80. Wallach: 'Misperceptions of Clausewitz's *On War*', p.230.

81. Theobald von Bethmann Hollweg: *Betrachtungen zum Weltkriege* (Berlin: Hobbing, 1919, 1921), Vol. 2, pp.7–9.

82. Friedrich von Bernhardi: *The War of the Future* (London: Hodder & Stoughton, 1920), pp.196–8.

83. Erich Ludendorff: *Kriegführung und Politik* (Berlin: Mittler, 1922), pp.104–5.

84. General Erich Ludendorff: *Der Totale Krieg*, transl. by A. S. Rappoport as *The Nation at War* (London: Hutchinson, 1936), pp.169–89.

85. Shaposhnikov, quoted in Raymond L. Garthoff: *Soviet Military Doctrine* (Glencoe, Ill.: The Free Press, 1953), p.11.

86. Freyer, Jünger and Forsthoff, quoted in Hans-Ulrich Wehler: 'Der Verfall der deutschen Kriegstheorie', in Ursula von Gersdorff (ed.): *Geschichte und Militärgeschichte* (Frankfurt/Main: Bernard Graefe, 1974), pp.291–6.

87. Quoted in Werner Hahlweg: 'Das Clausewitzbild einst und jetzt', in *Vom Kriege*, p.78.

88. Hans-Ulrich Wehler: 'Der Verfall der deutschen Kriegstheorie', in Gersdorff (ed.): *Geschichte und Militärgeschichte*, p.307.

89. Quoted in Jehuda Wallach: *Kriegstheorien* (Frankfurt/Main: Bernard und Graefe, 1972), pp.178f.
90. Quoted in *ibid.*, p.179.
91. Quoted in Panajotis Kondylis: *Theorie des Krieges: Clausewitz – Marx – Engels – Lenin* (Stuttgart: Klett-Cotta, 1988), pp.107f.
92. Wilhelm von Blume: *Strategie, ihre Aufgaben und Mittel* (3rd edn. of *Strategie, eine Studie*, Berlin: Ernst Siegfried Mittler und Sohn, 1912), pp.10f.
93. Wilhelm Groener: *Der Feldherr wider Willen* (Berlin: Mittler, 1930), p.164.
94. Ludwig Beck: *Studien* (Stuttgart: Koehler, 1955), pp.59f., 63.
95. Quoted in Klaus-Jürgen Müller: 'Clausewitz, Ludendorff and Beck', in Michael Handel (ed.): *Clausewitz and Modern Strategy* (London, Frank Cass, 1986), p.244.
96. Quoted in Müller: 'Clausewitz, Ludendorff and Beck', p.254.
97. Müller: 'Clausewitz, Ludendorff and Beck', p.248.
98. Bernard Brodie: *War and Politics* (London: Cassell, 1974, originally 1973), pp.13f.
99. Wallach: *The Dogma*, pp.249–310, *passim*.
100. Brodie: *War and Politics*, pp.38f.
101. *Ibid.*, particularly pp.9f., 188.
102. Zhang Yuan-Lin: 'Mao Zedong und Carl von Clausewitz', p.28.

4. Beyond Numbers: Genius, Morale, Concentration of Forces, Will and Friction

1. Hermann Cohen: *Von Kants Einfluss auf die deutsche Kultur, Marburger Universitätsrede* (Berlin: 1883), pp.31f. quoted in Hans Rothfels: *Carl von Clausewitz, Politik und Krieg* (Berlin: Dümmler, 1920), pp.23–5.
2. T.G. Otte: 'Educating Bellona: Carl von Clausewitz and Military Education', in Keith Nelson and Greg Kennedy (eds.): *Military Education: Past, Present, Future* (New York: Praeger, 2002).
3. *Vom Kriege*, I, 3, pp.232ff., 251, 283f.
4. Quoted in Ernst Hagemann: *Die deutsche Lehre vom Kriege: Von Berenhorst zu Clausewitz* (Berlin: E. S. Mittler & Sohn, 1940), p.14.
5. Quoted in Hagemann: *Die deutsche Lehre*, p.49.
6. *Der Krieg* (1815), quoted in Hagemann: *Die deutsche Lehre*, p.50; see also p.70.
7. Stephen J. Cimbala: *Clausewitz and Escalation: Classical Perspectives on Nuclear Strategy* (London: Frank Cass, 1991), pp.166–81.
8. Antulio J. Echevarria II: 'On the Brink of the Abyss: The Warrior Identity and German Military Thought before the Great War', *War and Society*, Vol. 13, No. 2 (October 1995), pp.29–31.
9. Clausewitz: 'Nachricht', in *Vom Kriege*, p.182.
10. 'Über die sr. königl . . . ', *Vom Kriege*, pp.1053f.

11. *Vom Kriege*, VI, 27–8, pp.810f., 814.
12. *Ibid.*, VIII, 9, p.1011.
13. Quoted in Ulrich Marwedel: *Carl von Clausewitz* (Boppard am Rhein: Harald Boldt Verlag, 1978), p.131.
14. Jay Luvaas: *The Education of an Army: British Military Thought, 1815–1940* (London: Cassell, 1965), pp.324f.
15. Major General Sir F. Maurice: *British Strategy: A Study of the Application of the Principles of War* (London: Constable, 1929), pp.27f.
16. Michael I. Handel: *Masters of War: Classical Strategic Thought* (2nd edn, London: Frank Cass, 1996), p.15.
17. *Vom Kriege*, III, 17, p.413.
18. 'Übersicht des Sr. königl. Hoheit . . . ' *Vom Kriege*, p.1070.
19. *Vom Kriege*, VIII, 4, pp.976f.; see also 'Über die sr. königl . . . ' (1810–12), *Vom Kriege*, pp.1049f.
20. *Vom Kriege*, VIII, 4 pp.976–7.
21. Napoleon: *Military Maxims*, in T. Phillips (ed.): *Roots of Strategy* (London: John Lane, the Bodley Head, 1943), p.236.
22. Quoted in Hagemann: *Die deutsche Lehre*, p.40.
23. Eberhard Kessel: 'Zur Genesis der Modernen Kriegslehre: Die Entstehungsgeschichte von Clausewitz's Buch "Vom Kriege"', *Wehrwissenschaftliche Rundschau*, 3rd year, No. 9 (1953), p.408.
24. Jomini: *Treatise on Grand Military Operations* (New York: D. van Nostrand, 1865), p.149.
25. Hubert Camon: *Clausewitz* (Paris: R. Chapelot, 1911), p.3.
26. *Vom Kriege*, III, 11, p.388.
27. 'Übersicht des sr. königl. Hoheit . . . ' in *Vom Kriege,* pp.1053f.
28. Marwedel: *Carl von Clausewitz*, pp.174f. Commandant J. Colin: *The Transformation of War*, transl. by L. H. R. Pope-Hennessy (London: Hugh Rees, 1912), pp.347, 350.
29. Großer Generalstab (ed.): *Moltkes Militärische Werke*, Vol. IV, *Kriegslehren*, Part 3 (Berlin: Ernst Mittler und Sohn, 1911), p.6.
30. Capitaine Georges Gilbert: *Essais de critique militaire* (Paris: La nouvelle revue, 1890), pp.1, 4.
31. Marshal [Ferdinand] Foch: *The Principles of War*, transl. by Hilaire Belloc (London: Chapman & Hall, 1918), p.3.
32. *Ibid.*, p.4.
33. *Ibid.*, pp.22, 25.
34. *Ibid.*, pp.34f.
35. *Ibid.*, pp.42f.
36. *Voennyi entsiklopedicheskii slovar* (Moscow, Voenizat, 1983), pp.397f.
37. Quoted in Liddell Hart: *The Ghost of Napoleon* (London: Faber and Faber, 1934?), pp.128f.
38. *Ibid.*, p.129.
39. *Ibid.*, p.139.
40. The US Marine Corps: *Warfighting* (New York: 1994), p.107, n.28,

quoted in Colin S. Gray: *Modern Strategy* (Oxford: Oxford University Press, 1999), p.96.

41. Eberhard Kessel (ed.): *Carl von Clausewitz: Strategie aus dem Jahr 1804 mit Zusätzen von 1808 und 1809* (2nd edn., Hamburg: Hanseatische Verlagsanstalt, 1943), pp.41f.

42. Azar Gat: *The Origins of Military Thought* (Oxford: Clarendon Press, 1989), pp.67–78.

43. Quoted in Peter Paret: *Clausewitz and the State* (Oxford: Clarendon Press, 1976), p.157.

44. Raymond Aron: *Clausewitz, Philosopher of War*, transl. by Christine Booker and Norman Stone (London: Routledge & Kegan Paul, 1976), p.120.

45. *Vom Kriege*, III, 3–5, pp.356–65.

46. 'Übersicht des sr. königl. Hoheit . . . ', *Vom Kriege*, p.1070.

47. *Ibid.*, IV, 10, pp.460–66.

48. Clausewitz: 'Nachricht', in *Vom Kriege*, p.183.

49. General Erich Ludendorff: *Der Totale Krieg*, transl. by A. S. Rappoport as *The Nation at War* (London: Hutchinson, 1936), p.20.

50. Commandant J. Colin: *The Transformations of War*, transl. by L. H. R. Pope-Hennessy (London: Hugh Rees, 1912), p.335.

51. Eugène Carrias: *La pensée militaire française* (Paris: Presses Universitaires de France, 1960), p.279.

52. Hubert Camon: *Clausewitz* (Paris: R. Chapelot, 1911), pp.1–3, 11.

53. Capitaine Georges Gilbert: *Essais de critique militaire* (Paris: Librairie de la Nouvelle Revue, 1890), pp.3, 20, 22, 278.

54. Georges Gilbert: *La guerre sud-africaine* (Paris: Berger-Levrault, 1902), pp.491f.

55. Quoted in Jack Snyder: *The Ideology of the Offensive: Military Decision Making and the Disasters of 1914* (Ithaca: Cornell University Press, 1984), p.80.

56. General Négrier: 'Quelques enseignements sur la guerre Russo-Japonaise', *Revue des deux mondes* (15 June 1906), p.34, quoted and translated in Douglas Porch: 'Clausewitz and the French, 1971–1914', in Michael Handel (ed.): *Clausewitz and Modern Strategy* (London, Frank Cass, 1986), p.299.

57. Porch: 'Clausewitz and the French', p.297.

58. *Ibid.*, p.299.

59. Foch: *The Principles of War*, pp.285–7.

60. General Palat: *La philosophie de la guerre d'après Clausewitz* (1st edn. Paris: Henri Charles-Lavauzelle, 1921; this edition Paris: Economica, 1998), p.266.

61. Lt General (Ret.) Horst von Metzsch: *Der einzige Schutz gegen die Niederlage: eine Fühlungnahme mit Clausewitz* (Breslau: Ferdinand Hirt, 1937), pp.17, 62.

62. Marwedel: *Carl von Clausewitz*, p.164.

63. Quoted in Emmanuel Terray: *Clausewitz* (Paris: Fayard, 1999), p.190.

64. *Vom Kriege*, I, 1, 2, pp.191f.

65. *Ibid.*, I, 1, p.193.

66. 'Übersicht des sr. königl.', *Vom Kriege*, p.1070.

67. *Vom Kriege*, IV, 11, p.469.

68. Manfred Rauchensteiner: 'Betrachtungen über die Wechselbeziehung von politiscjem Zweck und militärischem Ziel', in Clausewitz-Gesellschaft (ed.): *Freiheit ohne Krieg? Beiträge zur Strategie-Diskussion der Gegenwart im Spiegel der Theorie von Carl von Clausewitz* (Bonn: Dümmler, 1980), pp.63f.

69. Großer Generalstab (ed.): *Moltkes Militärische Werke*, Vol. IV, *Kriegslehren*, Part 3 (Berlin: Ernst Mittler und Sohn, 1911), p.6.

70. Antulio J. Echevarria II: 'On the Brink of the Abyss: The Warrior Identity and German Military Thought before the Great War', in *War and Society*, Vol. 13, No. 2 (October 1995), p.29.

71. Foch: *The Principles of War*, pp.282–7.

72. *Ibid.*, pp.287–90.

73. See my *The Bomb* (London: Longman, 2000), Ch.2.

74. Quoted in Karl Köhler: 'Jenseits von Clausewitz und Douhet', *Revue Militaire Générale*, No. 10 (December 1964), p.656.

75. Général André Beaufre: *Introduction à la stratégie* (Paris: Librairie Armand Colin, 1963), pp.15f.

76. *Vom Kriege*, III, 14, pp.401f.

77. *Ibid.*, IV, 7, p.443.

78. Colin: *The Transformations of War*, p.334.

79. Foch: *The Principles of War*, p.51.

80. *Ibid.*, p.97.

81. *Ibid.*, pp.292–8.

82. Jehuda Wallach: *Kriegstheorien: Ihre Entwicklung im 19. und 20. Jahrhundert* (Frankfurt/Main: Bernard & Graefe, 1972), pp.162f.

83. Quoted in Russell F. Weigley: *The American Way of War: A History of United States Military Strategy and Policy* (Bloomington: Indiana University Press, 1977), p.213.

84. *Vom Kriege*, I, 4, p.255; I, 5, p.257.

85. *Ibid.*, I, 7, pp.261f.

86. *Ibid.*, I, 7, p.263.

87. *Ibid.*, I, 8, p.265.

88. F. C. von Lossau: *Ideale der Kriegführung* (Berlin, 1836–9), Vol. III, p.340, quoted in Hagemann: *Die deutsche Lehre*, p.49.

89. *Vom Kriege*, II, 2, p.289.

90. William Safire: 'The Fog of War: Von Clausewitz Strikes Again', *International Herald Tribune* (19 November 2001).

91. *Vom Kriege*, I, 1, 21 and I, 3, pp.208, 234.

92. Moltke: *Militärische Werke*, Vol. II, 2, pp.208 and 293, quoted in Eberhard Kessel: *Moltke* (Stuttgart: K.F. Koehler, 1957), p.511.

5. The Defensive–Offensive Debate, the Annihilation Battle and Total War

1. Christopher Bassford: 'Landmarks in Defense Literature: *On War* by Carl von Clausewitz', *Defense Analysis*, Vol. 12, No. 2 (1996), p.268.
2. Quoted in Ernst Hagemann: *Die deutsche Lehre vom Kriege: Von Berenhorst zu Clausewitz* (Berlin: E. S. Mittler & Sohn, 1940), p.41.
3. *Ibid.*
4. *Ibid.*, p.53.
5. Carl von Clausewitz: 'Strategie aus dem Jahre 1804', in Werner Hahlweg (ed.): *Carl von Clausewitz: Verstreute kleine Schriften* (Osnabrück: Biblio Verlag, 1979), pp.25f.
6. Über die sr. königl . . .', *Vom Kriege*, pp.1075–8.
7. *Vom Kriege*, I, 1, 27, p.205.
8. *Ibid.*, VI, 1, pp.613f.
9. *Ibid.*, VI, 1, 2, p.615.
10. *Ibid.*, VII, 2, p.872.
11. *Ibid.*, VI, 5, p.633; also VII, 2, p.872.
12. *Ibid.*, VI, 8, pp.646ff.
13. *Ibid.*, VI, 9, p.665.
14. *Ibid.*, VI, 10, pp.670–81.
15. Ulrich Marwedel: *Carl von Clausewitz: Persönlichkeit und Wirkungsgeschichte seines Werkes bis 1918* (Boppard am Rhein: Harald Boldt Verlag, 1978), pp.167–72.
16. Großer Generalstab (ed.): *Moltkes Militärische Werke*, Vol. IV, *Kriegslehren*, Pt. 3, *Die Schlacht* (Berlin: Ernst Mittler & Sohn), pp.141f. Moltke used very similar language in his 'Wechselwirkung zwischen Offensive und Defensive', written 1869–85, see *ibid.*, pp.167–70.
17. *Ibid.*, pp.207f.
18. *Ibid.*, p.163.
19. Azar Gat: *Military Thought: The Nineteenth Century* (Oxford: Clarendon Press, 1992), p.67.
20. Quoted in Jack Snyder: *The Ideology of the Offensive: Military Decision Making and the Disasters of 1914* (Ithaca: Cornell University Press, 1984), p.139.
21. *On War*, Book VI. 2–8, pp.618–70.
22. Liet.-Gen. von Caemmerer: *The Development of Strategical Science during the Nineteenth Century* (trs. Karl von Donat; London: Hugh Rees, 1905), p.98.
23. Friedrich von Bernhardi: 'Clausewitz über Angriff und Verteidigung: Versuch einer Widerlegung', *Beiheft zum Militär-Wochenblatt*, No. 12 (Berlin: Ernst Siegfried Mittler & Sohn, 1911), pp.399–412, here pp.399, 411.
24. Alten: *Handbuch für Heer und Flotte* (1909–1914), Vol. 2, p.209, quoted in Marwedel: *Carl von Clausewitz*, pp.158f.
25. Quoted in Snyder: *The Ideology of the Offensive*, p.20.

26. Jehuda Wallach: *Kriegstheorien* (Frankfurt/Main: Bernard & Graefe, 1972), p.175.
27. Wallach: 'Misperceptions of Clausewitz's *On War* by the German Military', in Michael Handel (ed.): *Clausewitz and Modern Strategy* (London, Frank Cass, 1986), pp.224–6.
28. Introduction to Clausewitz: *Théorie de la grande guerre*, transl. de Vatry (1886), Vol. I, p.ix, quoted in Marwedel: *Carl von Clausewitz*, p.237.
29. Eugène Carrias: *La Pensée militaire française* (Paris: Presses Universitaires de France, 1960), p.255.
30. Quoted in Snyder: *The Ideology of the Offensive*, p.80.
31. Capitaine Georges Gilbert: *Essais de critique militaire* (Paris: Librairie de la Nouvelle Revue, 1890), pp.vi, 20.
32. *Ibid.*, pp.vi, 12f., 19, 22, 37, 46.
33. Alan Mitchell: *Victors and Vanquished: The German Influence in Army and Church in France after 1870* (Chapel Hill: University of North Carolina Press, 1984), pp.276f.
34. Gat: *The Nineteenth Century*, pp.122–5.
35. Derrécagaix: *La guerre moderne* (Paris: Librairie militaire de L. Baudouin, 1885), Vol. I, p.372.
36. Joseph Joffre: *Mémoires* (Paris: Plon, 1932), pp.32–3, quoted and translated in Douglas Porch: 'Clausewitz and the French, 1971–1914', in Michael Handel (ed.): *Clausewitz and Modern Strategy* (London, Frank Cass, 1986), pp.298f.
37. Colonel Maillard: *Éléments de la guerre* (Paris: Librairie militaire de L. Baudouin, 1891), p.435.
38. Raymond Aron: *Clausewitz, Philosopher of War*, transl. Christine Booker and Norman Stone (London: Routledge & Kegan Paul, 1976), p.247.
39. Bernard Brodie: *Strategy in the Missile Age* (2nd ed., Princeton, NJ: Princeton University Press, 1965), pp.52f.
40. Marshal [Ferdinand] Foch: *The Principles of War*, transl. Hilaire Belloc (London: Chapman & Hall, 1918), pp.283f.
41. Foch: *The Principles of War*, pp.281f.
42. Snyder: *The Ideology of the Offensive*, pp.16, 41.
43. *Ibid.*, pp.41–106.
44. Gat: *The Nineteenth Century*, p.131.
45. *Ibid.*, p.136.
46. Jean Jaurès: *L'Armée Nouvelle* (Paris: Publications Jules Rouffet, 1911), p.137.
47. Admiral Raoul Castex: *Strategic Theories*, transl. Eugenia Kiesling (Annapolis, Md: Naval Institute Press, 1994), pp.312, 344.
48. Russell F. Weigley: *The American Way of War: A History of United States Military Strategy and Policy* (Bloomington: Indiana University Press, 1977), p.213.
49. Brodie: *Strategy in the Missile Age*, p.178.
50. Major General Professor Rasin: 'Die Bedeutung von Clausewitz für die Entwicklung der Militärwissenschaft', *Militärwesen*, 2nd Year, No. 3 (May 1958), p.384.

51. Quoted in Raymond L. Garthoff: *Soviet Military Doctrine* (Glencoe, Ill.: Free Press, 1953), p.66.
52. Quoted in *ibid.*, p.66.
53. Quoted in David M. Glantz: *The Military Strategy of the Soviet Union – A History* (London: Frank Cass, 1992), p.280, n.45.
54. Quoted in Garthoff: *Soviet Military Doctrine*, p.67.
55. Quoted in Zhang Yuan-Lin: 'Mao Zedong und Carl von Clausewitz: Theorien des Krieges, Beziehung, Darstellung und Vergleich' (MS Ph.D., University of Mannheim, 1995), pp.167f.
56. Quoted in Azar Gat: *Fascist and Liberal Visions of War: Fuller, Liddell Hart, Douhet and Other Modernists* (Oxford: Clarendon Press, 1998), p.235.
57. Clausewitz: 'Strategie von 1804', in Werner Hahlweg (ed.): *Carl von Clausewitz: Verstreute kleine Schriften* (Osnabrück: Biblo Verlag, 1979), pp.20f.
58. 'Übersicht des Sr. Königl. Hoheit . . . ', in *Vom Kriege*, p.1070.
59. *Vom Kriege*, IV, 11, p.467.
60. *Ibid.*, VII, 4, p. 877.
61. *Ibid.*, VI, 25, pp.784–98.
62. *Ibid.*, VII, 7, p.883.
63. *Ibid.*, VI, 26, p.805.
64. Hagemann: *Die deutsche Lehre*, pp.54f.
65. Quoted in *ibid.*, p.105.
66. Friedrich Engels, letter of 30 March 1854, quoted in Panajotis Kondylis: *Theorie des Krieges: Clausewitz – Marx – Engels – Lenin* (Stuttgart: Klett-Cotta, 1988), p.150.
67. Marwedel: *Carl von Clausewitz*, p.131.
68. Großer Generalstab (ed.): *Moltkes Militärische Werke*, Vol. IV, *Kriegslehren*, Part 3 (Berlin: Ernst Mittler und Sohn, 1911), p.5.
69. Rudolf Stadelmann: *Moltke und der Staat* (Krefeld: Scherpe-Verlag, 1950), particularly p.369.
70. Hans-Ulrich Wehler: 'Der Verfall der deutschen Kriegstheorie', in Ursula von Gersdorff (ed.): *Geschichte und Militärgeschichte* (Franfkurt/Main: Bernard Graefe, 1974), p.287.
71. Helmuth von Moltke: *Aufzeichnungen, Briefe, Schriften, Reden* (Ebenhausen bei München: Wilhelm Langewiesche-Brandt, 1942), pp.337f.
72. *Ibid.*, pp.340f.
73. Quoted in Marwedel: *Carl von Clausewitz*, p.144.
74. Colmar von der Goltz: *The Nation in Arms* (London: Hugh Rees, 1906), p.470.
75. Colmar von der Goltz: *Jena to Eylau* (London: Kegan Paul, Trench, Trubner, 1913), p.74.
76. *Ibid.*, pp.70–3.
77. Von der Goltz: *The Nation in Arms*, p.77.
78. Colmar von der Goltz: *Kriegführung: Kurze Lehre ihrer wichtigsten Grundsätzxe und Formen* (Berlin: 1895), quoted in Kondylis: *Theorie des Krieges*, p.119.

79. Jehuda Wallach: *The Dogma of the Battle of Annihilation* (Westport, Ct.: Greenwood Press, 1986), p.41.
80. Jehuda Wallach: 'Misperceptions of Clausewitz's *On War* by the German Military', in Michael Handel (ed.): *Clausewitz and Modern Strategy* (London, Frank Cass, 1986), p.216.
81. *Ibid.*, p.217.
82. Clausewitz: 'Betrachtungen über einen künftigen Kriegsplan gegen Frankreich, 1830', in Werner Hahlweg (ed.): *Carl von Clausewitz: Verstreute Kleine Schriften* (Osnabrück: Biblio Verlag, 1979), pp.533–63.
83. Wallach: *The Dogma*, pp.69, 72.
84. Kondylis: *Theorie des Krieges*, pp.119f.
85. Snyder: *The Ideology of the Offensive*.
86. Manfried Rauchensteiner: 'Betrachtungen über die Wechselbeziehung von politischem Zweck und militärischem Ziel', in Clausewitz-Gesellschaft (ed.): *Freiheit ohne Krieg? Beiträge zur Strategie-Diskussion der Gegenwart im Spiegel der Theorie von Carl von Clausewitz* (Bonn: Dümmler, 1980), pp.65–7.
87. Gerhard Ritter: *The Schlieffen Plan*, transl. Andrew and Eva Wilson (London: Oswald Wolff, 1958), p.91.
88. *Vom Kriege*, I, 2, p.220; see also p.228.
89. *Ibid.*, VIII, 8, p.1004.
90. Hans Delbrück: *Friedrich, Napoleon, Moltke: Aeltere und neuere Strategie* (Im Anschluß an die Bernhardische Schrift: 'Delbrück, Friedrich der Große und Clausewitz', Berlin: Hermann Walther, 1892), pp.5f.; Hans Delbrück: 'Über die Verschiedenheit der Strategie Friedrichs und Napoleons', in Delbrück: *Historische und politische Aufsätze* (2nd edn., Berlin: Georg Stilke, 1908), pp.223–301.
91. Hans Delbrück: 'Falkenhayn und Ludendorff', *Preußische Jahrbücher*, Vol. 180, No. 2 (May 1920), pp.48f.
92. Hans Delbrück: *Die Strategie des Perikles* (Berlin: Georg Reimer, 1890), pp.2f.
93. *Ibid.*, p.5.
94. Scherff: *Delbrück und Berhardi* (1892), pp.7, 38, and Friedrich von Bernhardi: *Delbrück, Friedrich der Große und Clausewitz* (1892), quoted in Marwedel: *Carl von Clausewitz*, pp.160, 205.
95. Delbrück: *Die Strategie des Perikles*, pp.16f.
96. Hans Delbrück: 'Falkenhayn und Ludendorff', *Preußische Jahrbücher*, Vol. 180, No. 2 (May 1920), pp.48f.
97. Gert Buchheit: *Vernichtungs- oder Ermattungstrategie?* (Berlin: Paul Neff Verlag, 1942).
98. *Vom Kriege*, Book I, 2, p.220; Book VI, 8 and 9, pp.647–65.
99. For other writers, see for example W. Erfurth: *Der Vernichtungskrieg: Eine Studie über das Zusammenwirken getrennter Heeresteile* (Berlin: 1939), quoted in Kondylis: *Theorie des Krieges*, p.132, and see pp.132–6.
100. General Erich Ludendorff: *Der Totale Krieg*, transl. A. S. Rappoport as *The Nation at War* (London: Hutchinson, 1936), p.168.

101. Christopher Bassford: *Clausewitz in English* (Oxford: Oxford University Press, 1994), p.66.
102. Spenser Wilkinson, quoted in Luvaas: *The Education of an Army: British Military Thought*, pp.283f.
103. F. N. Maude: 'Introduction' in Carl von Clausewitz: *On War*, transl. by Colonel J. J. Graham (London: Kegan Paul, 1908), p.v.
104. J. F. C. Fuller: 'Introduction' in Joseph I. Greene (ed.): *The Living Thoughts of Clausewitz* (London: Cassel, 1945), pp.1f.
105. Major Stewart L. Murray: *The Reality of War: An Introduction to Clausewitz* (London: Hugh Rees, 1909), and *idem.*: *The Reality of War: A Companion to Clausewitz* (London: Hodder & Stoughton, 1914), pp.213f.
106. Admiral Sir Gerald Dickens: *Bombing and Strategy: The Fallacy of Total War* (London: Sampson Low, Marston & Co., 1947); and the reply by Capt. Robert H. McDonnell: 'Clausewitz and Strategic Bombing', *Air University Quarterly Review*, Vol. VI (Spring 1953), pp.43–54 (in defence of Clausewitz).
107. Capitaine Georges Gilbert: *Essais de critique militaire* (Paris: Librairie de la Nouvelle Revue, 1890), pp.10, 48.
108. Derrécagaix: *La guerre moderne* (Paris: Librairie militaire de L. Baudouin, 1885), pp.25–7.
109. Marshal [Ferdinand] Foch: *The Principles of War*, transl. by Hilaire Belloc (London: Chapman & Hall, 1918), p.299.
110. *Ibid.*, pp.295, 299.
111. Aron: *Clausewitz, Philosopher of War*, p.248.
112. Commandant J. Colin: *The Transformations of War*, transl. by L. H. R. Pope-Hennessy (London: Hugh Rees, 1912), p.335.
113. *Vom Kriege*, I, 2, p.217.
114. Christopher Bassford: 'John Keegan and the Grand Tradition of Thrashing Clausewitz', in *War in History*, Vol. 1, No.3 (1994); Basil H. Liddell Hart: *Paris or the Future of War* (London: Kegan Paul, Trench, Trubner, 1925), p.17; *idem.*: The Remaking of Modern Armies (London: John Murray, 1927), pp.103f.
115. Basil H. Liddell Hart: *The Ghost of Napoleon* (London: Faber and Faber, probably 1934), chapter-heading p.118, and this quotation p.120.
116. *Ibid.*, p.121.
117. *Ibid.*, p.122.
118. *Ibid.*, p.142.
119. Quoted in Michael Howard: 'The Influence of Clausewitz', in Michael Howard and Peter Paret (eds.): *Carl von Clausewitz: On War* (Princeton, NJ: Princeton University Press, 1976), p.42.
120. Field Manual 100–5 of 1 October 1939, quoted in Colonel Harry G. Summers Jr: *On Strategy: A Critical Analysis of the Vietnam War* (Novato, Ca.: Presidio, 1982), p.63.
121. FM 100–5, 19 February 1962, quoted in Summers: *On Strategy*, p.93.
122. FM 100–5, 6 September 1968, quoted in Summers: *On Strategy*, p.95.
123. Quoted in Raymond L. Garthoff: *Soviet Military Doctrine* (Glencoe, Ill.: Free Press, 1953), p.150.

124. Quoted in Manfred Backerra, 'Zur sowjetischen Militärdoktrin', *Beiträge zur Konfliktforschung*, Vol. 13, No. 1 (Jan. 1983), p.49.
125. Germany: Bundesarchiv, Militärisches Zwischenarchiv, MZP, VA-Strausberg/32657, pp.54–5, emphasis added.
126. Zhang Yuan-Lin: 'Mao Zedong und Carl von Clausewitz: Theorien des Krieges, Beziehung, Darstellung und Vergleich' (MS Ph.D., University of Mannheim, 1995), pp.28f.
127. Jan Philipp Reemtsma: 'Die Idee des Vernichtungskrieges: Clausewitz – Ludendorff – Hitler', in Hannes Heer & Klaus Naumann (eds): *Vernichtungskrieg: Verbrechen der Wehrmacht 1941–1944* (Hamburg: Hamburger Edition, 1995), pp.377–401.
128. Roger Chickering: 'Total War: The Use and Abuse of a Concept', in Manfred Boemeke, Roger Chickering and Stig Förster (eds): *Anticipating Total War: The German and American Experiences, 1871–1914* (Cambridge: Cambridge University Press, 1999), pp.15f.
129. Clausewitz: 'Bekenntnisdenkschrift', in Hans Rothfels (ed.): *Carl von Clausewitz: Politische Schriften und Briefe* (Munich: Drei Masken Verlag, 1922), p.750.
130. General Erich Ludendorff: *Der Totale Krieg*, transl. by A. S. Rappoport as *The Nation at War* (London: Hutchinson, 1936), pp.16, 23.
131. *Vom Kriege*, VIII, 2 and 3B, pp.954, 971–3, my italics.
132. Hans-Ulrich Wehler: 'Der Verfall der deutschen Kriegstheorie: Vom "absoluten" zum "Totalen" Krieg oder von Clausewitz zu Ludendorff', in Ursula von Gersdorff (ed.): *Geschichte und Militärgeschichte* (Frankfurt/Main: Bernard Graefe, 1974), pp.277f.
133. *Vom Kriege*, I, 1, 11, p.201.
134. Hew Strachan: 'On Total War and Modern War', *International History Review*, Vol. 22, No. 2 (June 2000), pp.241–343, 353f.; Wehler: 'Der Verfall der deutschen Kriegstheorie', p.288.
135. Commandant J. Colin: *The Transformations of War*, transl. by L. H. R. Pope-Hennessy (London: Hugh Rees, 1912), pp.342f.
136. A. S. Rappoport translated the German *total* as 'totalitarian' throughout, but wrongly so in my opinion.
137. General Erich Ludendorff: *Der Totale Krieg*, transl. by Rappoport, pp.11–16, 23, my italics.
138. *Les Guerres d'Enfer* (Paris: E. Sansot, 1915).
139. *La Guerre Nouvelle* (Paris: Lib. Armand Colin, 1916).
140. Paris: Nouvelle Librairie Nationale, 1918.
141. Strachan: 'On Total War and Modern War', pp.341–70.
142. Ludendorff: *The Nation at War*, pp.11–16, 23.
143. General Ludwig Beck: 'Die Lehre vom Totalen Kriege', in Günter Dill (ed.): *Clausewitz in Perspektive* (Frankfurt/Main: Ullstein, 1980), p.521.
144. *Vom Kriege*, Book I, Ch. 1.3, p.193.
145. *Ibid.*, pp.527–32.
146. Gerhard Ritter, letter to Beck of 4 November 1942, quoted in Müller: 'Clausewitz, Ludendorff and Beck', p.258, my italics.
147. Aron: *Clausewitz, Philosopher of War*, p.252.

148. Liddell Hart: *The Ghost of Napoleon*, p.120.
149. *Ibid.*, p.121.
150. *Ibid.*, p.144.
151. John Keegan: *War and Our World* (London: Hutchinson, 1998), pp.41–3; see also John Keegan: *A History of Warfare* (New York: Knopf, 1993).
152. Martin Shaw: *Dialectics of War: An Essay in the Social Theory of Total War and Peace* (London: Pluto Press, 1988), pp.16f., 61; and see Michael Howard: *Clausewitz* (Oxford: Oxford University Press, 1984), p.70.
153. James John Turner: *Just War Tradition and the Restraint of War: a Moral and Historical Inquiry* (Princeton, NJ: Princeton University Press, 1981), p.267.
154. Brodie: *Strategy in the Missile Age*, pp.37f.
155. *Ibid.*, p.315.
156. Jay Luvaas: 'Clausewitz, Fuller and Liddell Hart', in Michael I. Handel (ed.): *Clausewitz and Modern Strategy* (London: Frank Cass, 1986), p.197.

6. Taking Clausewitz Further: Corbett and Maritime Warfare, Mao and *Guerrilla*

1. Julian Corbett: *England in the Seven Years War: A Study in Combined Strategy*, 2 vols. (London: Longmans, Green & Co., 1907), Vol. I, p.6.
2. Corbett: *Some Principles of Maritime Strategy* (originally 1911, this reimpression London: Conway Maritime Press, 1972), p.134.
3. Quoted by B. McL. Ranft, Foreword to Corbett: *Some Principles*, p.ix.
4. Quoted in Donald M. Schurman: *Julian S. Corbett, 1854–1922: Historian of British Maritime Policy from Drake to Jellicoe* (London: Royal Historical Society, 1981), pp.53 f.
5. Corbett: *Some Principles*, p.14.
6. D. M. Schurman: *The Education of a Navy* (London: Cassell, 1965), p.183.
7. Corbett: *England in the Seven Years War*, Vol. I, pp.3f.
8. Corbett: *Some Principles*, pp.6f.
9. *Ibid.*, p.13.
10. *Ibid.*, p.14.
11. *Ibid.*, pp.16, 23.
12. *Ibid.*, p.18.
13. *Ibid.*, p.25.
14. *Ibid.*, pp.29f.
15. *Ibid.*, pp.28, 45, 50.
16. *Ibid.*, pp.51–4, my italics.
17. *Ibid.*, pp.54f.
18. *Ibid.*, p.58.
19. *Ibid.*, p.63.
20. *Ibid.*, pp.70f.
21. *Ibid.*, p.73.

22. *Ibid.*, p.87.
23. *Ibid.*, pp.89f.
24. *Ibid.*, p.94.
25. *Ibid.*, pp.157–62.
26. Jon Tetsuro Sumida: *Inventing Grand Strategy and Teaching Command: The Classic Works of Alfred Thayer Mahan Reconsidered* (Baltimore: Johns Hopkins University Press, 1997), p.113.
27. Joint Publication 1–02, *Department of Defense Dictionary of Military and Associated Terms* (1 Dec. 1989), p.212, quoted by General Larry D. Welch: 'Air Power in Low- and Midintensity Conflict', in Richard H. Shultz, Jr. and Robert L. Pfaltzgraff, Jr. (eds.): *The Future of Air Power in the Aftermath of the Gulf War* (Maxwell Air Force Base, Alabama: Air University Press, 1992), p.142.
28. *Vom Kriege*, VI, 26, p.799.
29. Full text and illustrations in Werner Hahlweg (ed.): *Carl von Clausewitz: Schriften – Aufsätze – Studien – Briefe* (Göttingen: Vandenhoeck & Ruprecht, 1966), pp.226–588.
30. Clausewitz: 'Bekenntnisdenkschrift', in Hans Rothfels (ed.): *Carl von Clausewitz: Politische Schriften und Briefe* (Munich: Drei Masken Verlag, 1922), pp.118f.
31. *Vom Kriege*, VI, 26, pp.799–806, particularly p.800.
32. *Ibid.*, VI, 6, pp.636–8.
33. *Ibid.*, VI, 26, pp.799f.
34. *Ibid.*, III, 17, p.413.
35. *Ibid.*, VI, 26, p.801.
36. *Ibid.*, III, 4, pp.359f.
37. *Ibid.*, VI, 26, p.803.
38. *Ibid.*, VI, 26, pp.803f.
39. *Ibid.*, VI, 26, p.805.
40. *Ibid.*, I, 2, p.221 and VI, 30, pp.833–6.
41. Both quoted in Werner Hahlweg: 'Lenin und Clausewitz', in Günter Dill (ed.): *Clausewitz in Perspektive* (Frankfurt/Main: Ullstein, 1980), pp.639f.
42. Christopher Bassford: *Clausewitz in English*, p.74.
43. T. E. Lawrence: *The Seven Pillars of Wisdom* (London: Jonathan Cape, 1973), pp.193, 196.
44. See T. E. Lawrence: 'Guerrilla warfare', entry in the *Encyclopedia Britannica* (14th edn, Chicago, Encyclopedia Britannica, 1929), Vol. 10, printed in Gérard Chaliand (ed.): *The Art of War in World History* (Berkeley: University of California Press, n.d.), pp.880–90.
45. Gray: *Modern Strategy* (Oxford: Oxford University Press, 1999).
46. W. I. Lenin: 'Über die Junius-Broschüre', in *Über Krieg, Armee und Militärwissenschaft*, Vol. I (Ost Berlin: Verlag des Ministeriums für Nationale Verteidigung, 1958), p.597.
47. Quoted in Zhang Yuan-Lin: 'Mao Zedong und Carl von Clausewitz: Theorien des Krieges, Beziehung, Darstellung und Vergleich' (MS Ph.D., University of Mannheim, 1995), p.31.

48. *Vom Kriege*, VI, 5, p.634, and Mao in Yuan-Lin: 'Mao und Clausewitz', p.32.
49. Quoted in Yuan-Lin: 'Mao und Clausewitz', p.221.
50. *Ibid.*, pp.226f.
51. Quoted in *ibid.*, pp.232f.
52. Quoted in *ibid.*, p.234.
53. Helmut Schmidt: *Menschen und Mächte* (Berlin: Goldmann/Siedler, 1987), p.359.
54. So did Clausewitz, of course, in Book VI of *On War*, which Haffner ignored in his argument.
55. Sebastian Haffner: 'Mao und Clausewitz', in Günter Dill (ed.): *Clausewitz in Perspektive* (Frankfurt/Main: Ullstein, 1980), pp.652–63.
56. Ernesto Che Guevara: 'Taktik und Strategie der lateinamerikanischen Revolution (1962)', in Günter Dill (ed.): *Clausewitz in Perspektive* (Frankfurt/Main: Ullstein, 1980), pp.664–77.
57. Tru'ông-Chinh: *The Resistance Will Win* (3rd edn, Hanoi: Foreign Languages Publishing House, 1966), pp.106f.

7. Clausewitz in the Nuclear Age

1. Once between 1923 and 1933 in a three-volume version, again in 1934 and in 1936, and, significantly, in 1941. Anon.: 'Über die Einfürungsartikel zum Buch C.v.Clausewitz' ' "Vom Kriege" ', *Voyennaya Mysl*, transl. in *Militärwesen*, 3rd year, No. 4 (July 1959), p.599.
2. C. N. Donnelly: *Heirs of Clausewitz: Change and Continuity in the Soviet War Machine* (London: Alliance Publishers for the Institute for European Defence and Strategic Studies, 1985).
3. Klemens Kowalke: 'Die Funktionale Bedeutung der Clausewitzschen Methodologie für die Formierung der sowjetischen internationalen Politik mit besonderer Berücksichtigung des Einsatzes militärischer Macht' (MS Dr phil. Mannheim, 1989/1990), p.79.
4. Stalin's letter was printed in *Bolshevik*, No. 3 (February 1947), pp.4–7. Quoted in Byron Dexter: 'Clausewitz and Soviet Strategy', *Foreign Affairs*, Vol. 29, No. 1 (Oct. 1950), pp.44f.
5. Jean-Christophe Romer: 'Quand l'Armée Rouge critiquait Clausewitz', in *Stratégique*, No. 33 (1st term, 1987), pp.97–111.
6. For examples, see Gat: *The Nineteenth Century*, p.238.
7. L. Leschtchinskii: *Bankrotstvo Voennoi Ideologii Germanskich Imperialistov* (Moscow: 1951), cited in Martin Kitchen: 'The Political History of Clausewitz', in *Journal of Strategic Studies*, Vol. 11, No. 1 (March 1988), pp.43f.
8. *Vom Kriege*, VI, 30, p.856.
9. *Ibid.*, II, 6, p.336.
10. H. Dinerstein: *War and the Soviet Union: Nuclear Weapons and the*

Revolution in Soviet Military and Political Thinking (New York: Atlantic Books/Stevens and Sons, 1959), pp.71–5.

11. Kowalke: 'Die Funktionale Bedeutung', p.167.
12. Quoted in Zhang Yuan-Lin: 'Mao Zedong und Carl von Clausewitz: Theorien des Krieges, Beziehung, Darstellung und Vergleich' (MS Ph.D. University of Mannheim, 1995), pp.33f.
13. Quoted in Yuan-Lin: 'Mao Zedong und Carl von Clausewitz', pp.34f.
14. J. A. Rasin: 'Die Bedeutung von Clausewitz für die Entwicklung der Militärwissenschaft', in *Militärwesen, Zeitschrift für Militärpolitik und Militärtheorie*, Vol. 2, No. 3 (May 1958), pp.377–92.
15. N. Talenski in *Strategie und Abrüstungspolitik der Sowjetunion: Ausgewählte Studien und Reden* (Frankfurt/Main: 1964), pp.174–82 and 183–89, quoted in Panajotis Kondylis: *Theorie des Krieges: Clausewitz – Marx – Engels – Lenin* (Stuttgart: Klett-Cotta, 1988), pp.286f.
16. Quoted in Kowalke: 'Die Funktionale Bedeutung', p.168.
17. V. D. Sokolovskiy: *Soviet Military Strategy*, transl. and ed. by Harriet Fast Scott (3rd edn, 1968, London: Macdonald and Jane's, 1975), pp.173–7.
18. Quoted in Yuan-Lin: 'Mao Zedong und Carl von Clausewitz', pp.223f.
19. N. Talensky: 'The "Absolute Weapon" and the Problem of Security', *International Affairs* (Moscow), No. 4 (April 1962), p.24.
20. Quoted in Kowalke: 'Die Funktionale Bedeutung', p.84.
21. Kowalke: 'Die Funktionale Bedeutung', p.83.
22. Quoted in Helmut Dahm: 'Die sowjetische Militär-Doktrin', in *Osteuropa*, Vol. 6 (1970), pp.394f.; E. Rybkin criticising Talenskij in 1965, cited in Klemens Kowalke: 'Die Funktionale Bedeutung', p.84.
23. V. Tsvetkov: 'Vydajushchiisja voennyi myslitel' XIX veka' in *Voenno-istoricheskii Zhurnal* (1964), No. 1, pp.47–9.
24. Kowalke: 'Die Funktionale Bedeutung', p.85.
25. W. Samkowoj: *Krieg und Koexistenz in sowjetischer Sicht* (Pfullingen: 1969), pp.52f., quoted in Kondylis: *Theorie des Krieges*, pp.287f.
26. Sokolovskiy: *Soviet Military Strategy*, pp.174, 177.
27. Kondylis: *Theorie des Krieges*, p.287.
28. Quoted in Helmut Dahm: 'Die sowjetische Militär-Doktrin', in *Osteuropa*, Vol. 6 (1970), p.395.
29. Examples in Kowalke: 'Die Funktionale Bedeutung', pp.86f.
30. V. Y. Savkin: *The Basic Principles of Operational Art and Tactics* (Washington, D.C.: 1972), pp.22–4, quoted in Martin Kitchen: 'The Political History of Clausewitz', in *Journal of Strategic Studies*, Vol. 11, No. 1 (March 1988), p.44.
31. A. Milovidow *et al.*: *The Philosophical Heritage of V. I. Lenin and Problems of Contemporary War* (Washington: 1974), pp.46f., quoted in Kondylis: *Theorie des Krieges*, pp.288f.
32. *Great Soviet Encyclopedia*, 3rd edn., 1970 (English translation New York: Macmillan, 1974), Vol. 5, p.652.

33. *Great Soviet Encyclopedia*, 3rd edn. 1973 (English translation New York: Macmillan, 1976), Vol. 12 p.114.
34. Quoted in Kowalke: 'Die Funktionale Bedeutung', p.88. See also Hans-Henning Schröder: 'Gorbachow und die Generäle', *Berichte des Bundes-instituts für ostwissenschaftliche und internationale Studien*, No. 45 (1987).
35. M. A. Gareyev: *M.V. Frunze: voennyj teoretik* (Moscow: Boennoe Izdateya'stvo, 1985), translated as *Frunze, Military Theorist* (London: Pergamon Brasseys, 1988), p.86.
36. *Kommunist*, Nos. 10, 15 (1986), quoted in Kowalke: 'Die Funktionale Bedeutung', p.89.
37. Quoted in Kowalke: 'Die Funktionale Bedeutung', p.89.
38. Kowalke: 'Die Funktionale Bedeutung', pp.89, 172f.; see also Daniil Proektor: 'O politike, Klauzevitse: s pobede v yadernoi voine', *Mezhdu-narodnaya zhizn*, No. 4 (1988).
39. Mikhail Gorbachov: *New Thinking for our Country and the World* (London: Collins, 1987), p.141.
40. Quoted in Kowalke: 'Die Funktionale Bedeutung', p.91.
41. Serebrjannikov: 'S uchetom real'nostej jadernogo veka' in *Kommunist Vooruzhennych Sil*, No. 3 (1987), pp.9–16.
42. Cited in Kowalke: 'Die Funktionale Bedeutung', p.89.
43. Russell F. Weigley: *The American Way of War: A History of United States Military Strategy and Poilicy* (London: 1973), p.365.
44. Raymond Aron: *Penser la guerre, Clausewitz*, Vol. II, *L'âge planétaire* (Paris: Gallimard, 1976), p.183
45. Peter R. Moody, Jr.: 'Clausewitz and the Fading Dialectic of War', *World Politics*, Vol. 31 (1978), pp.417–33.
46. Colin S. Gray and Keith Payne: 'Victory is possible', *Foreign Policy*, Vol. 39 (Summer 1980), pp.14–27.
47. Bruce R. Nardulli: 'Clausewitz and the Reorientation of Nuclear Strategy', *Journal of Strategic Studies*, Vol. 5, No. 4 (December 1982), p.506.
48. *Vom Kriege*, I, 1 p.210; General Ulrich de Maizière: 'Politische Führung und militärische Macht', in Clausewitz-Gesellschaft (ed.): *Freiheit ohne Krieg? Beiträge zur Strategie-Diskussion der Gegenwart im Spiegel der Theorie von Carl von Clausewitz* (Bonn: Dümmler, 1980), pp.97f.
49. Gerd Stamp: *Clausewitz im Atomzeitalter* (Wiesbaden: Rheinische Verlags-Anstalt, n.d.), pp.13f in editor's introduction.
50. Brodie: *Strategy in the Missile Age*, p.43.
51. *Ibid.*, p.55.
52. Maurice Leman: 'Clausewitz, prophète de l'Apocalypse', *Stratégie et Défense*, No. 6 (July 1980), pp.10–12.
53. Bernard Brodie: 'The Atomic Bomb and American Security', Memoran-dum No. 18, Yale Institute of International Studies, 1945, quoted in Bernard Brodie (ed.): *The Absolute Weapon* (New York: Harcourt, Brace, 1946), p.76.
54. Bernard Brodie: *War and Politics* (London: Cassell, 1974, originally 1973), p.421.

55. *Ibid.*, p.412.
56. Werner Gembruch: 'Die Faktoren "Technik" und "technische Entwicklung" in der Kriegslehre von Clausewitz', in Günter Dill (ed.): *Clausewitz in Perspektive* (Frankfurt/Main: Ullstein, 1980), pp.465–72.
57. 'Übersicht des Sr. königl. Hoheit . . . ', *Vom Kriege*, p.1048.
58. *Vom Kriege*, I, 1, 3, pp.192, 194.
59. *Ibid.*, I, 1, 4, p.194.
60. *Ibid.*, I, 1, 14, pp.203f.
61. *Ibid.*, I, 1, 11, p.200.
62. *Ibid.*, I, 2, pp.218f.
63. *Ibid.*, I, 2, pp.224f.
64. *Ibid.*, VIII, 2, p.952.
65. *Ibid.*, VIII, 3A, p.959.
66. Herman Kahn: *On Escalation: Metaphors and Scenarios* (London: Pall Mall Press, 1965), pp.3, 6f.
67. Stephen J. Cimbala: *Clausewitz and Escalation: Classical Perspectives on Nuclear Strategy* (London: Frank Cass, 1991), p.3. Cimbala finds fault with Clausewitz because he was not critical of the fact that Napoleon failed to have 'endgame' strategies (or in the jargon of the 1990s, 'exit strategies'); See *ibid.*, p.4.
68. *Ibid.*, pp.9–11; and see Stephen Cimbala: *Clausewitz and Chaos: Friction in War and Military Policy* (New York: Praeger, 2000).
69. Cimbala: *Clausewitz and Escalation*, pp.26–36.
70. *Ibid.*, pp.98–118.
71. *Ibid.*, pp.123–59.
72. *Ibid.*, p.165.
73. *Ibid.*, pp.185–94.
74. *Ibid.*, pp.200–203.
75. Anatol Rapoport, quoted in Günter Dill: 'Einleitung', in Günter Dill (ed.): *Clausewitz in Perspektive* (Frankfurt/Main: Ullstein, 1980), p.xxx.
76. Colin Gray: 'What had RAND wrought?', *Foreign Policy*, No. 4 (Fall 1971), pp.111ff.
77. Bernard Brodie: *War and Politics* (London: Cassell, 1974, originally 1973), pp.452f.
78. Exchange of letters between Howard and Gray in *International Security*, Vol. 6 (Summer 1981), pp.185–7. See also David Curtis Skaggs: 'Of Hawks, Doves and Owls: Michael Howard and Strategic Policy', *Armed Forces and Society*, Vol. 11, No. 4 (Summer 1985), pp.609–26.
79. Colin S. Gray: *Modern Strategy* (Oxford: Oxford University Press, 1999), pp.ix, xi.
80. *Ibid.*, pp.321–3.
81. *Ibid.*, pp.297, 316, my emphasis, and p.322.
82. Brodie: *War and Politics*, p.63.
83. William W. Kaufmann: 'Limited Warfare', in *id.* (ed.): *Military Policy and National Security* (Princeton, NJ: Princeton University Press, 1956), pp.132f.
84. *Ibid.*, p.136; Beatrice Heuser: 'Victory in a Nuclear War? A Comparison

of NATO and WTO War Aims and Strategies', *Contemporary European History*, Vol. 7, Part 3 (November 1998), pp.311–28.

85. Kaufmann: 'Limited Warfare', p.136.
86. *Ibid.*, pp.112–16.
87. Robert Endicott Osgood: *Limited War: The Challenge to American Strategy* (Chicago: University of Chicago Press, 1957), pp.20–22, 24.
88. *Ibid.*, pp.22f.
89. *Ibid.*, pp.22, 25f.
90. *Ibid.*, p.24.
91. *Ibid.*, p.26.
92. Robert Endicott Osgood: *Limited War Revisited* (Boulder, Colorado: Westview Press, 1979), p.3.
93. *Ibid.*, p.2.
94. *Ibid.*, p.3.
95. *Ibid.*, pp.4f., 7.
96. *Ibid.*, p.10.
97. *Ibid.*, p.11.
98. Morton Halperin: *Limited War in the Nuclear Age* (New York: John Wiley & Sons, 1963), pp.130f.
99. *Ibid.*, p.3.
100. *Ibid.*, p.2.
101. *Ibid.*, pp.1f.
102. Thomas Schelling: *The Strategy of Conflict* (New York: Oxford University Press, 1963), pp.260f.
103. *Ibid.*, p.5.
104. Brodie: *Strategy in the Missile Age*, Ch. 9: 'Limited War', pp.305–57, particularly p.308.
105. Robert Endicott Osgood: *Limited War Revisited* (Boulder, Colorado: Westview Press, 1979), p.3.
106. Brodie: *War and Politics*, pp.106f.
107. Brodie: *Strategy in the Missile Age*, p.312, note 2.
108. *Ibid.*, p.313f.
109. Raymond Aron: 'Zum Begriff einer politischen Strategie bei Clausewitz', in Clausewitz-Gesellschaft (ed.): *Freiheit ohne Krieg? Beiträge zur Strategie-Diskussion der Gegenwart im Spiegel der Theorie von Carl von Clausewitz* (Bonn: Dümmler, 1980), pp.42–51.
110. Colonel Harry G. Summers Jr: *On Strategy: A Critical Analysis of the Vietnam War* (Novato, Ca.: Presidio, 1982), p.6; Bernard Brodie: 'The continuing relevance of *On War*', in Michael Howard & Peter Paret (eds. and trs.): *On War* (Princeton: Princeton University Press, 1976), p.51.
111. Summers: *On Strategy*, p.4.
112. *Ibid.*, p.xiv.
113. *Ibid.*, p.2.
114. *Vom Kriege*, II, 2, p.279.
115. Summers: *On Strategy*, pp.44–7.
116. *Ibid.*, p.18.
117. *Ibid.*, pp.50f.

118. Henry Kissinger: *The White House Years* (Boston: Little, Brown & Co., 1979), pp.34f.
119. Cf. Edward J. Villacres and Christopher Bassford: 'Reclaiming the Clausewitzian Trinity', *Parameters* (Autumn 1995), also http://www.clausewitz.com/CWZHOME/Trinity/Trinity.htm.
120. Summers: *On Strategy*, p.6.
121. *Ibid.*, p.13.
122. *Ibid.*, p.19.
123. *Ibid.*, pp.35–7.
124. Lieutenant Colonel Albert Sidney Britt III: 'European Military Theory in the 18th and 19th centuries', *Supplementary Readings for Advanced Course in the History of Military Art* (West Point, New York: US Military Academy, 1971), p.10, quoted in Summers: *On Strategy*, p.3.
125. *Ibid.*, p.3.
126. Henry Kissinger: *The White House Years*, p.64.
127. FM 100–5 of September 1954, quoted in Summers: *On Strategy*, p.67.
128. FM 100–5 of February 1962, quoted in Summers, p.69.
129. Sir Robert Thompson: *Revolutionary War in World Strategy, 1945–1969* (New York: Crane, Russak, 1970), pp.16f.
130. FM 100–5 of September 1968, quoted by Summers: *On Strategy*, p.78.
131. *Vom Kriege*, VIII, 6, p.995.
132. General Fred C. Weyland, quoted in Summers: *On Strategy*, p.79.
133. *Ibid.*, pp.174, 177.
134. Quoted in *ibid.*, p.103.
135. *Ibid.*, pp.88–91, 129.
136. *Ibid.*, pp.141–50. For an account that negates Summers' line and complains that the US Army did not fight a counter-insurgency war but simply a conventional war, see Andrew F. Krepinevich: *The Army and Vietnam* (Baltimore: Johns Hopkins University Press, 1986).
137. Michael I. Handel: *Masters of War: Classical Strategic Thought* (2nd edn, London: Frank Cass, 1996), pp.9f.
138. Full text in Handel: *Masters of War*, pp. 188f. Here we of course have echoes of Summers' exaggerated emphasis of the Clausewitzian 'Trinity' of government, military and people, see Chapter 3.
139. *Ibid.*, pp.11f.

8. Clausewitz's Relevance in the Twenty-First Century

1. Richard H. Shultz, Jr.: 'Compellence and the Role of Air Power as a Political Instrument', in Richard H. Shultz, Jr. and Robert L. Pfaltzgraff, Jr. (eds.): *The Future of Air Power in the Aftermath of the Gulf War* (Maxwell Air Force Base, Alabama: Air University Press, 1992), pp.171–91.
2. Alastair Sedgwick: 'How to conquer markets', *Management Today* (January 1977), pp.59–61.
3. Jehuda Wallach: 'Misperceptions of Clausewitz's *On War* by the German

Military', in Michael Handel (ed.): *Clausewitz and Modern Strategy* (London, Frank Cass, 1986), p.217.

4. *Das Werk des Untersuchungsausschusses der Deutschen Verfassungsgebenden Nationalversammlung und des Deutschen Reichstages 1919–1928*, 4th series, vol. 3, p.224, quoted in Wallach: 'Misperceptions of Clausewitz's *On War*', p.231.

5. General Erich Ludendorff: *Der Totale Krieg*, transl. by A. S. Rappoport as *The Nation at War* (London: Hutchinson, 1936), p.12.

6. E.g., Lieutenant General Raymond B. Furlong: 'The Validity of Clausewitz's Judgments for the Sphere of Air and Space War', in Clausewitz-Gesellschaft (ed.): *Freiheit ohne Krieg? Beiträge zur Strategie-Diskussion der Gegenwart im Spiegel der Theorie von Carl von Clausewitz* (Bonn: Dümmler, 1980), pp.221–8.

7. See above, p.116.

8. Eberhard Kessel: 'Die doppelte Art des Krieges', *Wehrwissenschaftliche Rundschau*, 4th year, No. 7 (1957), p.301.

9. *Ibid.*, p.305.

10. *Ibid.*, p.307.

11. Johan Galtung: 'Das Kriegssystem', in K. J. Gantzel (ed.): *Herrschaft und Befreiung in der Weltgesellschaft* (Frankfurt/Main: 1975), pp.69–77, cited in Günter Dill: 'Einleitung', in Günter Dill (ed.): *Clausewitz in Perspektive* (Frankfurt/Main: Ullstein, 1980), pp.xiv, xxxvf.

12. Jan Willem Honig: 'Strategy in a Post-Clausewitzian Setting', in Gert de Nooy (ed.): *The Clausewitzian Dictum and the Future of Western Military Strategy* (The Hague: Kluwer Law International, 1997), p.113.

13. *Ibid.*, pp.109–21.

14. R. v. L.: *Handbuch für den Offizier zur Belehrung im Frieden und zum Gebrach im Felde*, Vol. II (Berlin: S. Reimer, 1818), p.12.

15. Liddell Hart: *Strategy: The Indirect Approach* (London: Faber and Faber, 1954), p.336.

16. Azar Gat: *Fascist and Liberal Visions of War: Fuller, Liddell Hart, Douhet and Other Modernists* (Oxford: Clarendon Press, 1998), p.307.

17. Edward N. Luttwack: 'Towards Post-Heroic Warfare', *Foreign Affairs*, Vol. 74 (May–June 1995).

18. Hew Strachan: 'On Total War and Modern War', *The International History Review*, Vol. 22, No. 2 (June 2000), pp.345–8.

19. *Vom Kriege*, III, 9, pp.379–84; 'Übersicht des Sr. Kgl. Hoheit . . . ', *Vom Kriege*, pp.1057, 1071.

20. *Vom Kriege*, I, 6, pp.258–60.

21. MacGregor Knox: 'Conclusion: Continuity and Revolution in the Making of Strategy', in Williamson Murray, MacGregor Knox and Alvin Bernstein (eds.): *The Making of Strategy: Rulers, States, and War* (Cambridge: Cambridge University Press, 1944), pp.641f.

22. See Chapter 7 above.

23. Gat: *The Origins*, pp.193f.

24. Gat: *The Nineteenth Century*, p.67.

25. Jomini: *Treatise on Grand Military Operations*, Vol. 1 (New York: D. Van Nostrand, 1865), p.xviii.
26. Jomini: *Treatise on Grand Military Operations*, Vol. 2 (New York: D. Van Nostrand, 1865), p.445.
27. Jomini: *Summary of the Art of War*, transl. by O. F. Winship and E. E. McLean (New York: Putnam & Co, 1854), p.325.
28. Clausewitz: *Werke*, Vol. IX, pp.3–106, particularly p.19, Hans Rothfels: *Carl von Clausewitz: Politik und Krieg* (Berlin: Dümmler, 1920), p.61.
29. Gat: *The Origins*, p.212.
30. *Vom Kriege*, VIII, 3B, p.973.
31. *Ibid.*, VIII, 3B, p.973.
32. *Ibid.*, II, 6, pp.340f.
33. Gat: *The Origins*, p.197.
34. Werner Hahlweg (ed.): *Carl von Clausewitz: Verstreute kleine Schriften* (Osnabrück: Biblio Verlag, 1979), pp.60f.
35. Clausewitz: 'Über abstrakte Grundsätze der Strategie' (1808), in Eberhard Kessel (ed.): *Carl von Clausewitz: Strategie aus dem Jahre 1804 mit Zusätzen von 1808 und 1809* (Hamburg: Hanseatische Verlagsanstalt, 1937), p.71.
36. *Vom Kriege*, pp.175f., quoted by Marie von Clausewitz.
37. *Ibid.*, III, 8, p.376.
38. *Ibid.*, II, 2, p.289.
39. *Ibid.*, II 5, p.315.
40. *Ibid.*, II, 4, p.307.
41. *Ibid.*, II, 5, pp.312–34.
42. *Ibid.*, II, 2, p.290.
43. *Ibid.*, Vorrede des Verfassers, p.185.
44. This political science jargon has nothing to do with Clausewitz's 'realism' nor with the philosophical notion that abstract concepts have an objective existence, nor indeed with 'realism' in its general definition of 'fidelity to nature in representation'. Instead, it merely denotes a school of political scientists who are excessively fixated on the quality and spirit of highly competitive inter-state relations in the second half of the nineteenth and the first half of the twentieth century.
45. Brodie: *War and Politics*, pp.223–75.
46. Quoted in Klemens Kowalke: 'Die Funktionale Bedeutung der Clause-witzschen Methodologie für die Formierung der sowjetischen internatio-nalen Politik mit besonderer Berücksichtigung des Einsatzes militärischer Macht' (MS Dr. phil. Mannheim, 1989/1990), p.85.
47. Martin van Creveld: 'The Eternal Clausewitz', in Michael Handel (ed.): *Clausewitz and Modern Strategy* (London, Frank Cass, 1986), p.35.
48. Martin van Creveld: *The Transformation of War* (New York: The Free Press, 1991), p.ix.
49. *Ibid.*, p.226.
50. V. D. Hanson: *The Western Way of War: Infantry Battle in Classical Greece* (New York: Knopf, 1989).

51. Chris Brown: *Understanding International Relations* (London: Macmillan, 1997), p.116.
52. Stephen J. Cimbala's *Clausewitz and Chaos: Friction in War and Military Policy* (New York: Praeger, December 2000) was published too late for me to be able to integrate its findings into this book.
53. Herman Bondi: 'The case for a nuclear defence policy', *Catalyst*, Vol. 1, No. 2 (Summer 1985).
54. T. G. Otte: 'Educating Bellona: Carl von Clausewitz and Military Education', in Keith Nelson and Greg Kennedy (eds.): *Military Education: Past, Present and Future* (New York: Praeger, 2001).

Index